DIS

Books by Arthur C. Clarke

Nonfiction
Interplanetary Flight
The Exploration of Space
The Exploration of the Moon
Going into Space
The Coast of Coral
The Making of a Moon
The Reefs of Taprobane
Voice Across the Sea
The Challenge of the Spaceship
The Challenge of the Sea
Profiles of the Future
Voices from the Sky
The Promise of Space
Report on Planet Three
The First Five Fathoms
Boy Beneath the Sea
Indian Ocean Adventure
Indian Ocean Treasure
The Treasure of the Great Reef

With the Editors of Life
Man and Space

With the Astronauts
First on the Moon

With Robert Silverberg
Into Space

With Chesley Bonestell
Beyond Jupiter

Fiction
Islands in the Sky
Prelude to Space
Against the Fall of Night
The Sands of Mars
Childhood's End
Expedition to Earth
Earthlight
Reach for Tomorrow
The City and the Stars
Tales from the "White Hart"
The Deep Range
The Other Side of the Sky
Across the Sea of Stars
A Fall of Moondust
From the Ocean, from the Stars
Tales of Ten Worlds
Dolphin Island
Glide Path
The Lion of Comarre
The Nine Billion Names of God
Prelude to Mars
The Lost Worlds of 2001
The Wind from the Sun
Rendezvous with Rama
Imperial Earth

With Stanley Kubrick
2001: A Space Odyssey

Arthur C. Clarke has also edited:

The Coming of the Space Age

Time Probe
Three for Tomorrow

THE VIEW FROM SERENDIP

THE VIEW FROM SERENDIP

Arthur C. Clarke

RANDOM HOUSE NEW YORK

Portions of this work have previously appeared in *Astronautics &
Aeronautics, Chicago Tribune Magazine, Fantasy & Science
Fiction, Penthouse, Playboy, Technology Review,* and *Vogue.*

*Grateful acknowledgment is made to the following for permission
to reprint previously published material:*
D.A.C. News: "Servant Problem—Oriental Style" and "How to Dig
Space" by Arthur C. Clarke. Copyright © 1962, 1964 by D.A.C.
News.
Daily Telegraph Colour Magazine: "Schoolmaster Satellite" by
Arthur C. Clarke. Copyright © 1971.
Doubleday and Company, Inc.: *The Frontiers of Knowledge:* The
Frank Nelson Doubleday Lectures at the Natural Museum of
History and Technology at the Smithsonian Institution, Wash-
ington, D.C. Copyright © 1975 by Doubleday & Co., Inc.
Harper & Row, Publishers, Inc.: Excerpts from *The Making of a
Moon,* by Arthur C. Clarke. Rev. ed. Copyright © 1957, 1958 by
Arthur C. Clarke; excerpts from *The Challenge of the Space
Ship* by Arthur C. Clarke. Copyright © 1959 by Arthur C.
Clarke; "Whether or not there is life on Mars now, there will be
by the end of this century" by Arthur C. Clarke, from *Mars and
the Mind of Man* by Ray Bradbury, Arthur C. Clarke, et al.
Copyright © 1973 by Harper & Row, Publishers, Inc.
Little, Brown and Company: Excerpts from *Shipwrecks and Ar-
chaeology: The Unharvested Sea* by Peter Throckmorton. Copy-
right © 1971 by Peter Throckmorton; excerpts from *First on
the Moon: The Astronauts' Own Story* by Neil Armstrong,
Michael Collins and Edwin E. Aldrin, Jr.; *Beyond Jupiter: The
Worlds of Tomorrow* by Chesley Bonestell and Arthur C. Clarke.
London Observer Colour Magazine: "The Sea of Sinbad" by Arthur
C. Clarke. Copyright © 1972.
The New York Times: "UFO's Explained" and "The UFO Con-
troversy in America" by Arthur C. Clarke (July 27, 1975). Copy-
right © 1975 by The New York Times Company; "The Times
and Time" by Arthur C. Clarke. Copyright © 1969 by The New
York Times Company.
Time, The Weekly News Magazine: "Beyond The Moon: No End"
by Arthur C. Clarke. Copyright © 1969 by Time, Inc.
Courtesy Time-Life Books Inc.: "Closing in on Life in Space" by
Arthur C. Clarke, from *Nature/Science Annual,* 1974 Edition.

Library of Congress Cataloging in Publication Data
Clarke, Arthur Charles, 1917-
The view from Serendip.
1. Clarke, Arthur Charles, 1917-   —Biography.
2. Authors, English—20th century—Biography. I. Title.
PR6005.L36Z52 1977   823'.9'14   77-5989
ISBN 0-394-41796-8

Manufactured in the United States of America
2 4 6 8 9 7 5 3
First Edition

To Nellie, for her 86th birthday

Contents

THE VIEW FROM SERENDIP

1

Concerning Serendipity

For the last twenty years, my life has been dominated by three S's—Space, Serendip, and the Sea.

Space came first, and indeed led to the others by a roundabout but now apparently inevitable route. In the late 1940's, I realised that the new techniques of free-diving, pioneered by Jacques Cousteau and Hans Hass, gave human beings a cheap and simple way of experiencing the "weightlessness" of space travel—or something very close to it. So I purchased flippers and face mask, clawed my way down into the depths of the local swimming pool, closed my eyes, and tried to imagine that I was in orbit. It worked fairly well—though just *how* well I do not expect to know until I take my first ride in the space shuttle, sometime in the early 1980's (which will also be my own not-quite-so-early sixties).

I very quickly discovered that water could provide more than pseudoweightlessness. In sufficiently large quantities, which were readily available over three-quarters of the earth's surface, it could also supply adventure, beauty, strangeness, wonder—as well as an almost infinite menagerie of strange creatures, which even the most exotic planets might find it difficult to surpass. So I left the swimming pool and rediscovered the sea.

It was a rediscovery because I had been born within a few hundred metres of the sea (at Minehead, Somerset) and had spent much of my youth in or beside the Bristol Channel. Though I have always been under the impression that I was an extremely poor swimmer, this could hardly

3

have been the case; I used to enjoy bathing in water so rough that spectators gathered along the sea wall, unable to decide whether I was waving or drowning. (Even in those days, I was a show-off, and it's much too late to do anything about it now.)

In my late teens, however, I abandoned the sea when I moved to London and a life of gentlemanly leisure in the civil service, only slightly distracted by the rantings of Hitler. Yet geographical separation was not the only reason why, for almost ten years, I actually avoided the water; the fact that I could no longer see without spectacles was even more important. It was no fun splashing around in a saline haze; why, I might even lose my sense of direction, and start heading out towards the Welsh coast, fifteen kilometres away. . . .

The invention of face masks changed all that; sea-going spectacles were now perfectly practical, and I could have twenty-twenty vision above *and* below water. In a remarkably short time, with considerable initial impetus from a disintegrating marriage, I was on my way to the mecca of all undersea explorers, the Australian Great Barrier Reef.

I can recommend the sea voyage from London to Sydney for anyone who wants both to read *The Lord of the Rings* and to write a novel of his/her own—though I cannot claim to have produced the whole of *The City and the Stars* aboard the P & O's venerable *Himalaya*. A good deal of my time was spent at the bottom of the swimming pool, hoping to improve the capacity of my lungs, since there was no guarantee that underwater breathing equipment would be available in the remoter regions of the Reef. (It wasn't.) Eventually I was able to stay submerged for almost four minutes, but then gave up out of consideration for the other passengers, who viewed my activities with increasing alarm.

It was, indeed, a foolish exercise, and I am perhaps lucky to have survived. Many divers have killed themselves by this trick of "hyperventilation"—flushing out the CO_2 in the lungs by taking deep, rapid breaths, thus inhibiting the normal breathing reflex and destroying, for several minutes, any further desire for oxygen. Hyperventilation can also produce permanent brain damage. My God, do you suppose...?

Anyway, in mid-December 1954 I arrived safely in Colombo, the largest city and main port of Ceylon. The *Himalaya* would be there for half a day, which allowed time for a fair amount of local sightseeing. Fortunately, I had already made contact with two people who had promised to show me around.

One of them easily tops my list of "The Most Unforgettable Characters I've Ever Met." Major R. Raven-Hart, O.B.E., was then a tall, distinguished-looking man of sixty-five with a straggly beard which gave him a distinct likeness to Conan Doyle's Professor Summerlee, F.R.S. (And if you don't know Professor Summerlee, gadfly and critic of the immortal Professor George Challenger, run—don't walk—to *The Lost World*, which still remains the finest example of pure science-fiction adventure ever written, dammit.)

Raven-Hart had been in the British *and* French armies in the First World War, and the Royal Air Force in the Second. In one of them (I'm not sure which) he had erected a radio station on top of the Great Pyramid. He had been given a medal by the Pope ("I know you're not a Christian, but it won't do you any harm," the pontiff had remarked) for rescuing a party of nuns in an enterprise that also involved T. E. Lawrence.

He was a remarkable linguist, reading a dozen languages and speaking five, and had published translations from Dutch, German, French, Swabian-German, and several Pacific tongues. Between the wars, he explored at least three continents by canoe, producing a series of books which were illustrated by his own drawings and photographs. He was also a sculptor, and once presented me with a small clay model of such embarrassing ambiguity that I was relieved when it mysteriously disappeared. Some idea of his range of interests may be gathered from the titles of his books: *Canoe Errant on the Nile; Down the Mississippi; Ceylon: History in Stone; Canoe in Australia; Germans in Dutch Ceylon; The Pybus Embassy to Kandy 1762; Where the Buddha Trod.* He also published a humorous book about the Royal Air Force: *R.A.F.'ing It.*

In many ways, Raven-Hart reminded me of that much more famous traveller, Sir Richard Burton. He had similar linguistic abilities, coupled with a love of exotic places, cultures, and mores; the notorious scandal that terminated Burton's Indian career would have fascinated him. And a later writer with whom he had much in common—including scientific interests—was Norman Douglas; if they ever met, which seems quite probable, they must have recognised each other as kindred souls.

Major Raven-Hart was over eighty when he died, but he was still translating and editing, having spent the last years of his life in Durban, exiled (apparently for financial reasons) from the land he loved. Certainly his fondness for Ceylon was apparent on that bright December afternoon in 1954, when we hired a cab and saw the sights of Colombo together. It was my first introduction to the fabulous East; the novelty has worn off, but the appeal remains unabated.

That same busy day I met my other Ceylon contact, zoologist-diver-artist Rodney Jonklaas, then assistant director of Colombo's magnificent zoo. Rodney suggested that, if I survived the perils of the Great Barrier Reef, I should come back to explore the seas of Ceylon. At that time I was making no long-range plans; *Jaws*, perhaps fortunately, was still twenty-one years in the future, but I was acutely aware of the reputation of Australian sharks. Ironically, I met very few on the Reef; it was not until I got to Ceylon that I encountered these magnificent creatures face to face. So I made no promises, either to Rodney or to the major, when I sailed out of Colombo harbour, heading for the adventures I later recorded in

The Coast of Coral, but by the time that volume was published, in 1956, I was already in the process of being Serendipidised.

It is curious how words that have always been around, lurking obscurely in the undergrowth, are suddenly discovered and become positively hackneyed. "Serendipity" does not even appear in the excellent 1936 Longmans, Green edition of Roget's *Thesaurus* which still serves mé well; yet I have come across it recently in articles on engineering and astronomy by authors who would certainly not consider themselves to be literary stylists. However, though the word is usually employed in its correct meaning—of something useful or valuable discovered by a happy chance—I have found that very few people know its actual origin.

Serendip (or Serendib) is one of the many ancient names of Ceylon; it derives from the Muslim traders' Sarandib. The Greeks and Romans called the island Taprobane; the indigenous name was Sri Lanka ("the Resplendent Land"), and since 1972 this has been its official designation, though the national airline is still Air Ceylon, and no one ever talks of Sri Lankian tea.

So much for Serendip. Now Serendipity.

The word was invented, or at least put on paper, by the essayist Horace Walpole in 1754—exactly two hundred years before I myself set foot in Serendip. According to the *Oxford English Dictionary,* Walpole told one of his numerous correspondents that "he had formed it upon the title of the fairy-tale *The Three Princes of Serendip,* the heroes of which were always making discoveries, by accident and sagacity, of things they were not in quest of." (Note that Walpole, like Churchill, was not afraid of ending a sentence with a preposition.)

Whether *The Three Princes* is a genuine folk tale, or whether Walpole made it up, I do not know. But the exotically melodious word "serendipity" obviously filled a gap in the English language, though it seems to have taken two centuries to get into general circulation. I doubt if I had ever heard of it, when my own life provided a perfect example of its application.

The Great Barrier Reef had been my objective when, almost by accident, I paused at Serendip for a single afternoon and saw it briefly through the eyes of two of its residents. Even when, a year later, I returned to write *The Reefs of Taprobane* (1957), I still did not know what I had discovered, for the excitements and distractions of the outside world (indeed, outside *worlds*) clouded my eyes. Though I became steadily more involved with the country, returning as a tourist at least once a year, it was not until the late sixties that I found it more and more painful to say good-bye, and felt completely happy nowhere else on earth.

Now, more than twenty years after I first set foot on the island, I have at last been able to arrange my life so that I need no longer leave. How I managed this—and, more important, *why*—is one of the themes of this book.

2

Dawn of the Space Age

Towards the end of 1957, my love affair with Ceylon and its seas was interrupted by a persistent beeping at 20.005 megahertz.

There are some traumatic experiences that remain frozen in time, so that every man remembers to the end of his life exactly where he was when he heard, for example, of the assassination of John F. Kennedy or the attack on Pearl Harbor. In my case, I can add the dropping of the first A-bomb and the orbiting of the first Sputnik.

In the small hours of 5 October, on the opening day of the Eighth International Astronautical Congress, I was awakened in my Barcelona hotel room by a phone call from London; the *Daily Express* wanted my comments on the new Russian satellite. Being as surprised as everyone else, I was hardly in a position to make any informed statement; after the initial exhilaration, I realised that I would have to hurry back to the typewriter. The previous month I had published *The Making of a Moon: The Story of the Earth Satellite Program,* and it was now obvious that the United States Navy's Vanguard—to which my book had been largely devoted—was not going to be first into space.

Actually, it was not even second, since Dr. Wernher von Braun's Army team, which had been waiting impatiently in the wings, put the first American satellite into orbit within eighty days of getting the go-ahead. But those people who still refer to Vanguard as a failure do not know what they are talking about; launch vehicles derived from it—at a cost which now seems ludicrous—have been the workhorses of NASA's scientific and applications programs for almost twenty years. Vanguard was one of the best bargains the hard-pressed American taxpayer ever got.

As soon as I returned from Barcelona, I hastily (in three days!) revised *The Making of a Moon* in time for a new printing in January 1958. Meanwhile, Sputnik II, carrying the dog Laika, had gone into orbit. Not only was this satellite six times as large as Sputnik I, and three *hundred* times the weight of Vanguard I, but it also carried a living creature, a clear indication of Russian hopes. Yet not even the most optimistic could have dreamed that Yuri Gagarin's flight was only four years ahead.

The Making of a Moon ended with these words:

> The tiny, swiftly moving satellites of today are only a beginning; soon they will be joined by more sedately travelling companions, swinging on wider orbits round the Earth. Many of these will not merely be visible to the naked eye—they will be spectacular objects, far enough out in space to miss Earth's shadow, and able to outshine any of the stars.
>
> Our sister planets average more than three moons apiece; Earth is a freak in possessing only one. No one can guess how many satellites our world will have, or how large they will be, when this century draws to its close. Yet even if they are nothing more substantial than plastic balloons covered with reflecting paint, they will change the pattern of the night sky.
>
> This is an awe-inspiring thought, that should bring humility as well as pride. For when the story of our age comes to be told, we will be remembered as the first of all men to put their sign among the stars.

Those words were actually written a year before any satellites had been launched. Sputnik II was sailing silently through the constellations, brighter than all but a few of the stars, when I added this paragraph to the revised edition of *The Making of a Moon.*

For the first time in history, something man-made has become celestial, has passed beyond the realm of mundane affairs into a region which once seemed reserved for the gods. "Men come and go, but Earth abides." So it will be with these new creations of our minds and hands. Some of the fragile metal spheres now lying on laboratory benches in Washington and Moscow will still be orbiting this planet ages hence, when the nations which launched them are no more than faint and distant echoes in the memory of Man.

Only fourteen years later (see Chapter 14) I returned to this same thought. And by then, it was already both absurdly unimaginative and hopelessly geocentric.

So the Space Age, of which I had dreamed since childhood but had never really expected to see in my lifetime, had now well and truly begun, and the next twelve years were to climax in Neil Armstrong's first footsteps on the moon. Since I had now put down more than tentative roots in Ceylon and was sharing a small house with a large stockpile of diving gear, this led to a somewhat schizophrenic existence. I tried to straighten matters out with an essay, "Which Way Is Up?" (*The Challenge of the Spaceship,* 1960; reprinted in *Report on Planet Three,* 1972). After pointing out the numerous parallels between the sea and space, I concluded:

In the final analysis, I went undersea because I liked it there, because it opened up to me a new, strange world as fantastic and magical as the one which Alice discovered behind the looking glass. And perhaps I did it because, after hearing people call me a space-travel expert for twenty years, I felt I was getting into a rut. As Hollywood stars know very well, it is fatal to become typed; if you want to progress, to continue your mental and emotional growth, every so often you must surprise yourself (and your friends) by changing the pattern of your life and interests. . . . When there's nothing more to be said about you, you're already dead.

The 1960 edition of *Challenge of the Spaceship* (a title inspired by Arnold Toynbee's "challenge and response" theory of history) ended with a brief postscript, which again demonstrates how difficult it is to predict the rate of progress in an exploding field :

> Across the gulf of centuries, the blind smile of Homer is turned upon our age. Along the echoing corridors of time, the roar of the rockets merges now with the creak of the wind-taut rigging. For somewhere in the world today, still unconscious of his destiny, walks the boy who will be the first Odysseus of the Age of Space. . . .

Only three years after Sputnik I that still seemed a highly optimistic remark. But in 1960, Neil Armstrong, far from being a boy, was already *thirty* years old!

Fortunately, I was completely unspecific about the particular odyssey I had in mind. And it has just occurred to me that, right now, Dave Bowman is about ten years old . . . .

3

Servant Problem— Oriental Style*

Yet despite the distractions of space in the late fifties, I managed to keep a toe dipped in the sea. Using the photographs my partner, Mike Wilson, had accumulated in our decade of underwater activity together, I assembled two juvenile books: *Boy Beneath the Sea* (1958) and *The First Five Fathoms* (1960). Both still enjoy modest sales, and I am eternally grateful to Jacques Cousteau for taking time out from all his simultaneous careers to write a charming preface to *F.F.F.*

At that time, Mike, his wife, Elizabeth, and I were sharing a small bungalow only a short distance from the sea in Colpetty, a suburb of Colombo which had known better days. I am able to date the period of residence very accurately, for I have vivid memories of watching the moon from the garden through my Questar telescope at 21 hours 02m. 23s. on 13 September 1959—the instant Luna 2 impacted. (I saw nothing; nor did anyone else, though most of the telescopes on the sublunar hemisphere must then have been focused on the Mare Imbrium.)

---

* In these egalitarian and anticolonial times, this chapter may upset some readers. So let me point out that the employment situation described is almost totally unrelated to race, but only to ability, education—and of course luck (or Karma . . .).

Anyone who has read my fiction (especially *Childhood's End* or the short story *Reunion*) will know *my* views on racism. Nor am I ashamed of the fact that, in a country where the unemployment figure approaches 25 percent, I directly or indirectly support at least fifty people, many of whom might otherwise be literally starving.

A more down-to-earth record of our life in Rhineland Place (the Colombo Municipal Council has not yet marked the historic building with a plaque) is provided by an article I wrote a couple of years later, when we had moved to a somewhat larger house in Cinnamon Gardens. Here it is, just as it appeared in 1961, with no alterations. In fact, surprisingly little *has* altered—see the postscript following.

During the past five years, by the most modest estimate, at least twenty house-servants of assorted sizes, colours, religions, and races have entered my employment, and all but three (at the time of going to press) have left it again. This gives me a background of experience which few Western householders can match, and many may envy. Perhaps they will not envy it quite so much if I relate some of the episodes of my domestic saga.

When Mike Wilson and I arrived in Ceylon, back in 1956, one of the first orders of business was to find a reliable houseboy who would purchase our food, cook our meals, and generally look after our modest establishment. In the East, of course, a servant is not a luxury but a necessity. The time and money saved by a houseboy who knows the best places to go shopping can more than pay his wages; the local storekeepers would have made quite a killing had we two scrutable Occidentals dealt with them directly. (Not that this was beneath our dignity, as it would have been with many pukka sahibs. We didn't give a hoot for the prestige of the slightly moribund raj.)

We are still very grateful to the Ceylonese friend who sent us our first boy; despite certain flaws which will be mentioned later, Appuhamy was a pearl beyond price. He turned up, neat and nervous, early one morning at the hotel which we had made our base until we could find a suitable flat, and modestly proffered a document which in the years to come was to be very familiar to us. This was his "Servant's Pocket Register."

A government-issued booklet, printed in Sinhalese, English, and Tamil, the register looks very much like a passport and serves a similar purpose. It was invented by the British for self-protection back in the 1870's, and every servant was once supposed to have one. This rule, like many of the other good ideas of the departed imperialists, is no longer rigidly enforced; but a wise employer still views with suspicion any applicant without a register. It is a means of identification in case of trouble, and when you sign on a new servant you take possession of his register as kind of informal hostage. As it contains references from all his previous employers, it is one of his most valuable possessions and he won't leave suddenly without it.

The register starts off in a businesslike manner with a set of fingerprints, then follows a pretty comprehensive inventory. Here are the items of information that have to be listed: Full Name; Father's Name; Village; District; Race, Caste, and Religion; Trade or Profession; Property; Age; Height; Make; Complexion; Nose; Mouth; Eyes— colour and shape; Hair—colour and texture, and how arranged; Teeth; Description of hair on face; Usual expression of countenance. Any permanent peculiarity, scars, or marks on the head, face, hands, feet, body, or limbs? Any further peculiarity? General residence; Present residence and date. Married or single and what family and where living?

Some of these headings must be very baffling to a simple villager. "Make" means "build," and I have yet to see an entry under "Any further peculiarity," though heaven knows there have been cases when I could think of some. "Usual expression of countenance" seems a little unfair, for who is at his best when being fingerprinted and inspected for scars or marks? In the case of poor Appuhamy, the single word "Morose" had been written here by the registrar who had made the entries more than

twenty years before. It is, incidentally, no longer necessary to give caste, but this is a tricky subject we had
better leave alone.

After the inventory, like the space for visas in a passport, follow twenty pages in which particulars of employment can be listed. In some cases, needless to say, this
is nothing like enough, and one may be presented with a
whole bound volume of registers containing the interesting
comments of dozens and dozens of employers. In each
case, the record gives the name and address of the employer, the capacity in which the servant was hired and
the amount of pay, the date of engagement and discharge,
and—most important of all—the cause of discharge and
the testimonial (if any).

In Appuhamy's case, the testimonials were remarkably
good. They began with one that concluded regretfully:
"He leaves me to try and get better pay. Present salary
20 rupees [that is, four dollars] a month." This was,
admittedly, written quite a few years ago; we were considerably more generous, offering 100 rupees as a start
with the promise of more later if he proved to be satisfactory. This, I might say, was a very fair figure, and in
practice Appuhamy did much better than this. Possibly
much, *much* better; we often suspected so. Not that we
doubted his honesty, but it is the accepted practice to
make a little commission on the daily marketing, and
there must have been quite a flow of "surplus" food from
our kitchen to Appuhamy's own household.

For our "boy" was over forty when he started to work
for us and already had nine children. He was an alert,
dark little Sinhalese who barely topped the five-foot
mark, and looked rather like Popeye without the muscles.
(Or L'il Abner's Ma, now that I come to think of it.)
While other servants came and went, Appuhamy survived
in our eccentric but not very exacting household. Altogether he was with us for over four years—a record
which, alas, I do not imagine will ever be broken.

Every morning he presented us with the accounts for

the previous day, neatly drawn up in a cashbook. If you ever saw the movie *Elephant Walk*, which was shot in Ceylon, you may remember that Elizabeth Taylor's implacable cook (also called Appuhamy) baffled her by keeping the kitchen accounts in Sinhalese. We did not have to cope with this beautiful but impenetrable script, as one of our Appuhamy's numerous sons did the job very neatly in English. I cannot help reproducing one of the pages from this document, even though I have already quoted it in the book of our adventures, *The Reefs of Taprobane*. If you divide the figures by five, you have a rough idea of the values in dollars and cents. U.S. supermarkets, please copy.

To Cash Rupees 10

One pinapple	90
5 pairs	50
1 lb Coly flour	1.50
Soup vegitable	40
6 grapes fruit	1.50
½ lb kindneys	75
1½ lb pork chups	2.25
1 tin corn mutton	2.75
Spents	10.55
I want's balance	55

Appuhamy was an excellent cook and we became quite attached to him. We think that he was also fond of us; he should have been, for we let him have very much his own way. In the matter of Nona, for instance. . . .

I have already mentioned that Appuhamy had a wife and nine children (ten by the end of his first year with us). However, all was not well at home, which was in a rather tough quarter of Colombo. One morning Appuhamy arrived on his ancient bike in a state of some excitement to tell us that his next-door neighbor had hacked his wife (the neighbor's, not Appuhamy's) into several pieces. But such things can happen in any neighborhood, and it was not *this* that worried Appuhamy. His trouble was that his wife, very unreasonably, objected to sharing the house with his girl friend, Nona.

Appuhamy's sensible suggestion was that, since we had plenty of room and could do with some extra help, he and Nona should take up residence with us. Nona would help with the laundry and housework, and we would pay her a small wage. Perhaps I should explain that in Ceylon all houses have separate and self-contained servants' quarters, so that such an arrangement would be quite in order from the point of view of everyone except Mrs. Appuhamy.

Indeed, it worked admirably until Nona became conspicuously pear-shaped and departed for a short vacation in her home village. She returned to duty a few weeks later with a lovely baby boy, of whom Appuhamy was extremely proud. We, however, had reservations. To our suspicious minds, little Bevis bore a remarkable resemblance, not to Appuhamy, but to our junior houseboy, Jinadasa, whom I had better introduce.

Jinadasa was about nineteen, and a lad with ambition. You will be able to visualise him exactly when I tell you that he looked like an Asian Elvis Presley. Since rock-'n'-roll hit Ceylon, a lot of Sinhalese youths have started to look like Elvis Presley, done medium-rare.

To this day, we have never decided whether Jinadasa was extremely simple or extremely clever. One day, for instance, he produced a roll of Ektachrome film which he said he'd exposed in his camera and asked if I would process it for him. I was not too suprised about the camera, but I was surprised by the color film, which represented about a week's wages for Jinadasa. (Like all our servants, before and after, he was overpaid.) However, we had a considerable stock of Ektachrome for our own use in the darkroom, and I wondered. . . .

Consumed with curiosity, I processed the film. (The pictures were lousy.) As I expected, the emulsion number on the edge was the same as for our stock; presumably it had not occurred to Jinadasa that one could identify a roll of film. But when I charged him with theft, he was quite indignant and completely unabashed. "Then where did you get it?" I asked skeptically. Without a moment's hesitation, he named a local photographer. Hot on the trail, I took Jinadasa and film to the shop, where my suspect immediately engaged the proprietor in such fluent and confident Sinhalese that I retired baffled, giving him the benefit of the doubt. Not that there could have been much doubt, since we had brought our film stock direct from the States, and the chance that anyone else in Ceylon had Ektachrome with the same emulsion number was infinitesimal. By this time, however, I was so lost in admiration of Jinadasa's sheer cheek at stealing my film and *giving it back to me to process* that I let him get away with it.

Perhaps encouraged by this supineness, Jinadasa then made a much more ambitious move. One morning there arrived through the post this amazing letter; I have it in front of me at the moment and am copying it word for word:

Dear Sir,
    I desire to product a Sinhalese flim, but I have no money. Can you give me money to product a flim. If you give me money. When I finished my flim, I will returning your money. Please help me.
                                Thank you
                        Your obedient servant
                                Jinadasa

When I had digested this, I called for Jinadasa. For once, he looked a little sheepish.

"Do you know what it costs to make a flim—I mean a film?" I asked him. "How much did you expect me to lend you?"

He calculated for a few seconds; one could almost see the wheels going round.

"About three lakhs, master."

As a matter of interest, three lakhs (300,000 rupees,

or, say, $60,000) is just about what it *does* cost to produce
a Sinhalese film; we happen to know now, because we
are making one at the moment. But we didn't know in
those days, and I still wonder if Jinadasa's figure was
just a hopeful guess.

In any event, I did not leave him in suspense, but
quickly informed him that he had overestimated both my
gullibility and my bank balance. We never heard any more
about his film-making ambitions, but to keep temptation
out of the way we put a new lock on the darkroom door.

Jinadasa finally departed when it was obvious that he
could not save three lakhs from his pay, and that even if
he did his lunch break would not be long enough to allow
him to produce anything except a Grade B quickie. The
separation was quite amicable, and we wished Jinadasa
luck in his attempts to get financial backing. We also
gave him a reference which would not discourage future
employers—and that, I might say, required considerable
ingenuity. For now it is time to discuss the fascinating
section of the Servant's Pocket Register devoted to the
cause of discharge—and the employer's testimonial.

Every time we had a change of staff, I had to wrestle
with this—twice, in fact. The incoming man's document
had to be subjected to a most searching form of literary
criticism, and I had to compose an epitaph for the one
who was departing. As far as the last problem is concerned, I can say at once that some of my finest creative
work has been doomed to blush unseen in the pocket
registers of sundry Ceylonese cooks and houseboys. For
unless he has actually burned the house down, one does
not like to damn a man—perhaps for thirty or forty
years to come—by condemning him outright. A bad
reference, in a country with a high level of unemployment,
is a sentence of economic death.

On the other hand, it is dishonest—or at least unfair—
to send out an ex-servant with a testimonial that he does
not deserve. It is also liable to backfire; sooner or later a
friend will accost you in great indignation with: "Look

here—I took on this fellow Gunadasa because *you* said he was honest, and he's just left with all my silver."

It is therefore necessary, in sheer self-defence, to acquire the technique of composing—and interpreting—the ambiguous reference. This hints at the truth without actually saying it; I can best explain by giving a few examples, with free translations.

"Good, plain cook." (You'll need plenty of magnesia.)

"Appears honest." (We could never prove anything, but you've been warned.)

"Needs more experience." (But not at the expense of *our* stomachs.)

"Leaves to better his prospects." (Hopes to find a bigger sucker.)

"Not overfond of work." (Time-lapse photography might reveal signs of movement.)

"Works well under supervision." (We'll be glad to lend you our rhinoceros-hide whip.)

"Leaves at own request." (He just beat us to it.)

"Left because of disagreement with other servants." (The charge was reduced to manslaughter.)

And, most ominous of all, the two little words: "Means well."

With Jinadasa's departure, there was a succession of faces. Some are blurred together, others mercifully forgotten; I will mention only the highlights.

The midnight shriek with which Banda gave his notice woke up the whole neighborhood. That was really a piece of bad luck for which no one (except possibly Mike) was to blame. Here is the improbable sequence of events.

Banda slept in the garage, among the aqualungs, air compressors, and other tools of our trade. In the middle of the night, a strange noise disturbed him; it was merely, we deduced later, something sparking inside a battery charger we had left running. Nervously, Banda switched on the flashlight he kept beside his bed and sprayed the beam around the garage.

It was unfortunate that Mike has a macabre and somewhat juvenile sense of humor. A few years before he had bought, in a New York joke shop, a couple of imitation shrunken heads—very realistic. When the novelty wore off, he hung them up in the garage, and had quite forgotten them. But Banda's flashlight found them again, and he swore that one was twirling its straggly moustache. Just how a head sans body could do this he was unable to explain, during the short while it took him to pack his possessions.

Some time after Banda came Gamini, who was a very pretty boy with soulful eyes and an enigmatic smile. He was not much good in the kitchen and seemed to spend most of his time combing his hair and arranging flowers in the living room. When some of our friends started looking at us a little oddly, we decided that Gamini should decorate another household. We passed him on to a fashion-designer friend in whose elegant establishment, we were sure, he would be very happy.

Then there was Piyadasa, who was caught red-handed stealing some marked rupee bills and shipped back to his village in tears and disgrace. And Sirisena, who suddenly refused to work but stood for hours quite motionless, looking as if someone had bewitched him. As was perhaps the case, for charms and spells are still endemic in Ceylon, and only the other day we had to call in an exorcist to deal with an enchanted air conditioner. We never found the cure for Sirisena and grew a little tired of sharing the house with a waxwork, so he had to leave.

Undoubtedly the most spectacular departure was Ari's. At the time, we owned a small but completely fearless monkey named Liz, who was quite good-tempered—as monkeys go. Soon after the arrival of Ari, however, she became very highly strung and sensitive to unintended slights; we were sure that Ari was teasing her but were unable to prove it. All that was certain was that she hated him and flew into paroxysms of fury whenever he approached her cage.

Liz was quite an escape artist, and she bided her time.

When she was ready, she staged a jail break and took off after Ari, waving her arms and chittering with rage. After one glance at his approaching two-foot-high nemesis, Ari fled from the house and took refuge with the neighbors. Liz went in after him. For about ten seconds there was one of those breathless pauses that you see in the movies when you're waiting for the bridge to blow up, or the avalanche to start moving. Then the house next door erupted. Our shrieking neighbors—we had no idea there were so many of them—fled out of every door and several windows. There was no alternative but for me to go in, with many apologies, and capture the simian Clytemnestra. When I found Liz, she was dancing round on a dressing table waving a powder puff and leaving a trail of overturned scent bottles. She surrendered without a fight, but it was a long time before we were on speaking terms with our neighbors. Muslims do not appreciate strange men and stranger animals invading the boudoirs of their womenfolk.

The upshot of this episode was that we had to make a straight choice between Ari and Liz. We did not hesitate.

Through all these comings and goings, these alarms and excursions, Appuhamy survived. He was, as you may have gathered, a durable character and would probably have been with us yet if there had not been a drastic change in the status of our household. It suddenly (as these things happen) ceased to be a purely bachelor establishment. Mike's wife, Elizabeth, joined us, and the situation depicted in *Elephant Walk* reproduced itself with uncanny accuracy.

Appuhamy resented his demotion from the head of the house. He also objected to Elizabeth's gimlet-eyed audit of his accounts, and her constant querying of the price of "coly flour" and "pork chups." The cost of running our establishment dropped sharply, but only at the price of domestic peace.

It would be unfair to both parties to suggest that Elizabeth's arrival drove Appuhamy to drink; most of the time he was quite an abstemious and even a straitlaced person. (I can still remember his indignation when a well-known American author dropped round to see us, leaving the house redolent of hashish. "Very bad, master," grumbled Appuhamy.) Yet there were occasions when the whiskey bottle showed an extraordinary rate of evaporation, and Appuhamy waited at table with an expression very far from the "Morose" given in his Servant's Pocket Register.

The climax came on a certain Christmas Eve, when we returned from one party to find that Appuhamy had been having another. He was lying flat on the couch in the main living room, which reeked of arrack. When he saw us arrive, he attempted to rise but couldn't quite make it. There was a brief exchange of courtesies with Elizabeth, which she refused to translate but obviously never quite forgave. Since it was, after all, Christmas, we registered no more than a mild protest at the time, but I date the final decline and fall of Appuhamy from that moment.

Just before the parting of the ways, I was diffidently presented with the following note:

Sir,

As you are already aware, it is with regret that I will be leaving your services as cook from tomorrow after a period of eight years faithful service. During this period perhaps you will remember that I had a nasty fall and dislocated my arm. I am therefore unable to undertake any heavy work.

Sir I am a familied man and will greatly appreciate it if you can see your way to grant me a bonus for loyal service and since also because of my broken arm for which act of kindness I shall ever be grateful.

Yours obediently,
Appuhamy

Despite its faint hint of litigation (the reference was to an incident where for weeks Appuhamy refused to get proper treatment for a fractured wrist, insisting on herbal medicine, until we finally dragged him to our own doctor), we found this letter very touching. I may say at once that we gave Appuhamy a substantial (our friends said ridiculous) bonus, with which he was well satisfied. He still sends us greeting cards on feast days, and last Christmas bicycled round with the glad news that Mrs. Appuhamy had produced her eleventh child, and Nona her second.

Appuhamy's letter, however, contained one curious error that has often puzzled me, the phrase "*eight* years faithful service." We won't argue about the "faithful," but the fact of the matter is that Appuhamy was with us just over four years. I sometimes wonder if the arithmetic in his account book was equally erroneous.

Or did it *really* seem like eight years to him?

———

Looking back on this piece from no less than fifteen years later, I am astonished to see how little has changed. Most of the alterations have been quantitative; instead of three servants, there are, I have just been alarmed to discover, thirteen—and Appuhamy is back! Indeed, his absence lasted little more than a year; he returned when Elizabeth Wilson left to set up her own household, and my establishment reverted to its bachelor status.

At least, for most of the time. The various ladies who have occasionally resided here knew better than to pry too deeply into the workings of the kitchen, and—perhaps fortunately—none of them spoke Sinhalese. Thus, when Appuhamy muttered negative endearments under his sometimes alcoholic breath, only the other servants understood, and they were sensible enough to provide no translation.

I have just realised that, in a few months' time, it will have been twenty years since I met Appuhamy, and some kind of souvenir would seem to be in order. We have a genuine though guarded affection for each other, and he is largely responsible for the fact that I am two kilos overweight. (Well, three.)

But he has already been provided with a brand-new kitchen, a bicycle, and three assistants. And it seems a little late in the day for a complimentary vasectomy.

4

The Scent of Treasure

In the early 1960's the entire pattern of my life was totally disrupted by an event as unexpected, though hardly as significant, as the launching of Sputnik I. Once again, and in a manner I would never have dared to imagine, the sea began to dominate my existence.

A few years earlier, Mike Wilson had discovered virgin underwater territory on a dangerous reef ten kilometres off the south coast of Ceylon. No one had ever dived there before, and indeed operations were possible only during the two or three months between the end of the northeast monsoon and the onset of the southwest. But the site was attractive because of its spectacular rock formations, fish that had never been scared by spearguns, and a forty-metre-high lighthouse which, thanks to the hospitality of Trinity House's resident superintendent, could provide spartan and novel living quarters.

Mike, Rodney Jonklaas, and I travelled to the Great Basses Reef in the spring of 1959. One of our objectives, I now recall with some embarrassment, was to obtain dramatic colour photographs of Arthur Clarke disporting himself with sharks—and relaxing afterwards with a well-known brand of whiskey. This would have been slightly dishonest, since I happen to hate spirits, being instead addicted to Bristol Cream sherry and various sweet liqueurs referred to disparagingly as "cough syrups" by the serious drinkers of my acquaintance.

Exhibit A, the bottle of C - n - d - - n C - - b, had been tenderly carried by Mike all the way from the United States. Its contents, however, had failed to survive the journey. No matter; tea of the correct dilution would give exactly the same result, photographically if not physiologically. But as it happened, we never used it for its intended purpose, though the sharks were cooperative enough. Possibly I had belated qualms of conscience: at that time I had done only one commercial—a live TV spot for the London *Daily Express* when, with great enterprise, it intercepted the first close-ups of the moon's surface from Luna 9, and published them ahead of the Russians.

The 1959 Great Basses trip provided the most exciting —and occasionally hazardous—diving I had ever known and resulted in a short book, *Indian Ocean Adventure* (1961). I expected that this would be the end of the matter, for the expedition had been a strenuous business which I had no desire to repeat. Besides, I was very busy at the typewriter, being involved in a series of articles for *Playboy,* which became *Profiles of the Future,* as well as *two* novels: *A Fall of Moondust* and the third (and still not final) version of *Glide Path.*

So I was more than happy to let Mike go off to the Reef without me in the spring of 1961. This time he planned to shoot a movie, and indeed produced some beautiful sequences which are still sitting somewhere in a can, slowly rotting. *Boy Beneath the Sea* was abandoned for bigger game; later that year, while everything was still fresh in my mind, I explained how it all happened.

The oceans of the world are full of sunken treasure, but only once or twice in a generation does anyone stumble across it. This is how it happened with us.

For five years Mike Wilson and I have been diving around the coast of Ceylon, filming the unbelievably beautiful scenery beneath the Indian Ocean and collecting still photographs for our books. Our expeditions have taken us to some out-of-the-way places, of which the most inaccessible is a dangerous reef stretching for many kilometers along the southern coast of the island, far out at sea. The Great Basses Reef (the name is derived from the Portuguese *Baxos,* meaning a shoal or reef) can be approached only during a few weeks in every year, between the monsoons; for most of the time, it is a foaming death trap upon which countless ships have run aground. Today it is marked by a tall granite lighthouse, whose construction a century ago was a brilliant feat of engineering; it was built in Scotland and shipped to Ceylon! The four-man crew that maintains it has one of the loneliest jobs in the world; to the south there is nothing but the empty Indian Ocean, stretching for ten thousand kilometers to the Antarctic ice.

For several years we have visited the Great Basses Reef with our cameras and aqualungs, and its inhabitants have grown to know us well. In particular, we have made friends with a family of three large groupers, whom we've christened Sinbad, Aladdin, and Ali Baba. They remember us from year to year and are quite cooperative, though inclined to be rough when hungry. In return for food, they will go through a considerable repertory of acts in front of our cameras; you will find photos of them performing in our book *Indian Ocean Adventure.*

Last year Mike decided to produce an underwater film fantasy in which a small boy would meet these fish, and be escorted by them through their submarine empire. Our (human) actor was a thirteen-year-old American, Mark Smith, and assisting Mike was Mark's friend Bobby Kriegel, fourteen. Both boys were champion swimmers, and their trusting parents (members of the American official community in Ceylon) allowed Mike to cart them off to the Reef with about a ton of diving gear, air compressors, cameras, and canned food. (There is always a possibility of being marooned for a few weeks if the weather turns bad and the relief boat can't get to the lighthouse.)

The filming went very well, though a hungry Sinbad once swallowed Mark's arm up to the elbow until he was persuaded to let go, and *Boy Beneath the Sea* will be a masterpiece when Mike finally gets round to finishing the sound track. However, the movie no longer has top priority, thanks to one memorable day when the water was not clear enough for photography.

Determined not to waste their time in this unique undersea fairyland, Mike and the two boys went for a long swim over the reef, exploring territory they'd never visited before. First they tried to get near a school of porpoise, but failed, though they could hear the animals squeaking in the water around them. Almost a kilometre from the lighthouse, as he was passing over a ledge about five metres below the surface, Mike glanced down and spotted something he'd dreamed of finding ever since he'd taken up skin-diving. Lying in the open, as bright and shiny as if it had been dropped there only yesterday, was a beautiful little bronze cannon—certain proof that a wreck was close at hand.

Unable to believe his eyes, Mike dived down and tried to move the cannon. It was real enough and too heavy to be lifted. Signalling to the boys to start searching, Mike began a careful investigation of the seabed.

Almost at once a second small cannon was found; then, a few metres further on, were other unmistakable traces of marine disaster, encrusted by the coral of centuries. These later finds were too deep to be reached by skin-diving, so the three explorers hurried back to the lighthouse and collected their aqualungs, as well as the large inflated tire they used as a raft. Back at the site, they were examining the coral mounds when Bobby saw something gleaming brightly on the seabed. He pointed it out to Mike—whose yell of "Silver!" was audible even through the aqualung's mouthpiece.

Scattered over the seabed were thousands of coins, some loose and others aggregated into lumps. The lumps were cemented to the reef with coral, and at first sight looked like ordinary rock; no one would have given them a second glance. It took hard work with a crowbar to prize them loose, but eventually, after several visits, Mike and the boys were able to salvage more than fifty kilograms of pure silver, as well as the two bronze cannon, some copper bars and pins, and a handful of lead shot.

When one of the lumps was split open, the coins inside were found to be in perfect condition; indeed, they appear to have come straight from the mint. They are covered with elegant Persian characters, and all bear the Muslim date 1113, which is 1702 in our chronology. They are rupees, minted in Surat, India, during the reign of the Mogul Emperor Aurangzeb (1618–1707). One lump, judging by its weight, contains exactly one thousand rupees, and is still in the shape of the original bag, which held together long enough for the sea to cement the coins; there are even traces of fabric adhering to them. This particularly beautiful specimen, weighing about ten kilograms, we have donated to the Smithsonian Institution, where it will be displayed in an underwater archaeology exhibit.

There is no space here to tell of how the boys heroically kept the secret even from their parents, of our anxious consultations with lawyers, tax experts and customs officials, of guarded enquiries to the British Museum and London coin dealers. . . . All this is still continuing, and will be the subject of a full-length book—when we have returned to the Reef with proper salvage equipment. There are many questions still to be answered, and some that we may never answer. In 1702 the Dutch were in power in Ceylon; if the ship belonged to them, there may be records of her cargo, and we will know what else to seek. On the other hand, it is quite possible that she was an Arab or Chinese trader who had collected her "silk money" from Surat and never reached home. There could have been no survivors on this cruel reef far out at sea.

The several thousand dollars' worth of coins so far

salvaged may be only a fraction of the wealth still lying on the seabed. But even if we find no more, we have already been involved in a unique and exciting adventure— the discovery of the first treasure ever found in the Indian Ocean. It is sitting in my office now, locked up in a massive wooden chest near me as I type these words. When I lift the lid, I can smell a curious, metallic tang, which brings back vivid memories of the sea and of the spray-drenched rocks glittering beneath the equatorial sun.

It is the scent of treasure.

5

The Stars in Their Courses

Take my word for it, since you are unlikely ever to suffer the experience yourself: the finding of sunken treasure has a most unsettling effect upon the routine of everyday life. Though we did not expect to make our fortunes from the wreck (in fact the result was quite the opposite) there was no harm in daydreaming. After all, where there was so much silver, there might also be gold. . . .

In any event, there should be book royalties and TV rights, so Mike and I signed contracts with Harper & Row and the BBC which, we hoped, would at least cover the expense of a return to the Reef in a properly equipped boat of our own. And because time was of the essence— sooner or later, the secret would get out—we planned to hit the Great Basses at the very next opportunity, which would be in March 1962.

But while we were consulting lawyers, overhauling equipment, buying boats, and coping with the endless details of any expedition, things were starting to happen in Space. Not, this time, the new-fangled Space of countdowns and launch schedules, but the immemorial one of the astronomers or, rather, the astrologers.

The opening of 1962 was viewed with great alarm by millions of Asians, because there would be a total eclipse of the sun on 5 February, and no less than five planets would be in that same area of the sky when the corona blazed out in all its glory. For some reason, this was supposed to be a bad thing rather than a magnificent spectacle, and the most dire disasters were predicted. Needless to say, these could be averted—or at least minimised—if substantial payments were made to the experts in such matters.

I did my best to calm the panic, and to strike a blow for modern science, by a series of articles in the local press. The final one, "Common Sense About the Planets," appeared in the *Ceylon Sunday Times* for 28 January 1962; and my conclusion that there was nothing to worry about came within an ace of being the very last public statement I ever made.

Here is the piece concerned, which I now reread with a certain amused embarrassment. Remember, please, that it was written for an intelligent but scientifically naïve audience—probably 50 percent of whom believed implicitly in astrology. Which, I suspect, is the percentage in Los Angeles right now.

Much of the alarm that has been caused by the planetary conjunction of 5 February would vanish if people were aware of one simple fact. Contrary to many statements that have been made, such an occurrence is not particularly rare. Similar groupings of the planets take place about once every hundred years; in fact, there were two in the nineteenth century, and there will be two in this. The common belief that such a conjunction is unique in human history—or indeed in the history of the universe—is totally incorrect. Consequently, all the arguments and predictions based on this assumption are so much nonsense, and no one need be scared by them.

On the morning of 5 February, the sun, moon, and five of the planets will actually occupy quite a broad area of the sky—the width, in fact, of more than thirty moons. Because they are so close to the sun, all the planets are lost in the glare of the daylight sky, so no one can see the rather attractive grouping they will make. However, on the morning of the fifth there is a total eclipse of the sun, and for a maximum of four minutes there will be almost complete darkness along a very narrow band in mid-Pacific. The few fortunate observers in this band (which crosses New Guinea and the Solomon Islands) will be able to pick out the planets during the brief duration of totality, if they can tear their eyes away from the glory of the solar corona.

A Belgian astronomer, Jean Meeus, has now calculated just how often this sort of close grouping can occur, and has given his results in the December 1961 issue of the excellent journal *Sky and Telescope*. The answers are most interesting, and rather surprising. There are people around now whose lives will span three such celestial gatherings—so let us hear no more about their exceptional rarity!

Here, for anyone who wishes to do some historical investigating, are the dates when sun, moon, and the five brightest planets lie within an arc of 30 degrees, or width, of one sign of the Zodiac: 15 August 1007; 15 September 1186; 10 December 1284; 6 January 1285; 30 October 1483; 5 February 1524; 11 September 1624; 10 December 1662; 2 April 1821; 30 April 1821; 5 February 1962; 5 May 2000; 8 September 2040; 2 November 2100.

You will note that two of these conjunctions occur in pairs, within a month of each other. Four cases are associated with partial or annular eclipses of the sun, and the event of September 1186 is particularly interesting. The seven bodies then formed a tighter group than they will this February—and they were all on the same (east) side of the sun, not scattered on both sides of it as they will be this February.

The conjunction of 5 February 1524 would have been exceptionally beautiful, had it not been lost in the glare of the sun, for the three planets, Mars, Jupiter, and Saturn, then formed a tiny equilateral triangle covering an area not much bigger than the full moon. As the seven bodies were in the signs of Aquarius and Pisces, a general flood was predicted—but in many places February was the driest month of the year!

Since catastrophes and disasters are so common, I have no doubt that diligent research will turn up events that could be associated with some of the above groupings; I shall be quite disappointed if something spectacular did not happen in the busy month of April 1821! But I think that the list given by Dr. Meeus is enough to convince anyone not beyond the reach of reason that there is nothing to worry about this February. During the course of human history we have survived perhaps fifty such planetary groupings, some of them much closer and more remarkable than the one now approaching. If we do not survive many more, it will be through our own folly—not because of the heavenly bodies.

It is encouraging to see that many political figures in this country, and in the East generally, have publicly decried the present astrological hysteria. All those in a position of authority should do so, and if they fail they are guilty of a disservice to their community. For the danger with such predictions is that they may be "self-fulfilling"; if enough people believe them, they may come true. In India, for example, business on the stock exchange has slowed down because investors are fearful of the future. It needs no stellar influence to cause a financial crisis—if sufficient businessmen believe that there *will* be a financial crisis! The same argument applies in all other fields of human activity, for no enterprise will go ahead if those involved are convinced of its failure.

Nevertheless, I am going to predict one catastrophe as a result of the planetary configuration now approaching. It *will* result in disaster—to the astrologers. For after this, they will not have a leg to stand on; and I do not think that even their dupes will listen to them when they reply: "Ah, but you just wait until 5 May 2000!"

Well—if disaster *did* strike the local astrologers, they quickly recovered. It took me a good deal longer.

Shortly after the inauspicious day had arrived and uneventfully departed, I was shopping in Pettah, the Colombo bazaar, when I failed to notice a low archway and used my head like one of those large steel balls popular on demolition projects. Apart from the pain, there were no immediate effects; after I had stopped cursing, I got on my scooter and rode home. But that evening I started vomiting, and in the course of the next few days became so weak that I was unable to move and was carried, flopping round like a sack of jello, to a local nursing home.

Here, in due course, rival specialists diagnosed either (1) a spinal injury causing damage to the nervous system or (2) polio. The latter was a great surprise, but at the time a polio epidemic was raging in Ceylon, and Dr. Sabin himself had just visited the island. When I was again able to talk, I pointed out to the polio faction that it was indeed an odd coincidence that my collapse occurred a few hours after a violent head blow. The virus fanciers refused to budge.

Whatever the diagnosis, my recovery was uneventful, though protracted. During the course of several months, I evolved from an amoeba to something resembling an invertebrate. Within a year I was able to walk without sticks, and eventually without a limp. Today I am quite unaware of any remaining physical handicap, though in fact my left arm is useless for heavy work. But it is perfectly adequate for its main function, which is operating the shift, lock, and qwerty keys of my electric typewriter.

At the time, I was too ill to realise how ill I was, and my main concern was our forthcoming expedition to the Great Basses Reef. I can still remember pleading with the doctor to get me out of bed by next month, so that I could join the divers; the ominous word "manic" appears on my medical report for that time. Still, better manic than depressive, and I don't recall being particularly downcast. I might well have been, had I known that another year would pass before I could visit the silver wreck—and *that* would be pushing my luck.

The ominous year 1962 did, however, bring me one major compensation: I was still lying helpless in bed when the news was brought to me that I had won the UNESCO-administered Kalinga Prize for Science Writing. This annual one-thousand-pound award, donated by the Indian industrialist and statesman B. Patnaik, was a tremendous boost not only to my morale but to my bank balance, sadly depleted by recent events. I shall always be grateful to Mr. Patnaik for his timely assistance, and I was grieved to hear that he had been rounded up in Mrs. Gandhi's 1975 purge.[1]

The one-year delay did have some major advantages. It meant that our preparations were now only half-baked, instead of quarter-baked; and just as we were preparing to set off in April 1963 we acquired some invaluable volunteer help from a totally unexpected source. The famous diver-photographer-archaeologist Peter Throckmorton, who had discovered the oldest (1300 B.C.) wreck in the world, wrote from Greece asking if he could be of any assistance. He could, indeed; we sent him an airline ticket by return mail.

Peter's account of the expedition will be found in chapter 2 of his *Shipwrecks and Archaeology* (Boston: Atlantic—Little, Brown, 1971) and his description of the material salvaged gives an excellent idea of the site:

> There were the remains of a pair of matched flint-lock pistols. The wooden stock of one was like new, but the barrel had been dissolved by corrosion. There were the forearm of a musket; a pewter decanter stopper; bits of broken blue-and-white china and other shards; a bronze pestle; a gold-washed brass earring with green glass pendants; bits of green

---

1. The 1977 election, however, restored him to power.

glass bottles; fragments of bone; the brass buttplate of a musket; pistol and musket balls; and a silver-plated copper salver. Pieces of the coconut fiber bags which had held the coins, probably a thousand to the bag, were mixed in the solid mass like straw in bricks. Coins were everywhere, loose if the bag had rotted away before the coral had cemented them, or in the shape of the original bags.

For my own part, I published two books about the adventure—a short juvenile, *Indian Ocean Treasure,* and the fuller account, *The Treasure of the Great Reef* (New York: Harper & Row, both 1964). In 1974 the latter was reissued by Ballantine Books with an extensive collection of color plates and a new epilogue, "Ten Years Later."

Writing that epilogue was a curious and rather unsettling experience, for by then I was looking back on the greatest adventure of my life from the far side of Apollo, *2001,* and much else. Thanks to this new viewpoint, which I shall leave for a more appropriate moment (Chapter 13), I was to recognise the advent of serendipity again.

Looking for treasure, I had found something more important.

6

How to Dig Space

By the end of 1963 I had not merely managed to survive; I was starting to prevail. The treasure books were finished and the spell of the Great Basses Reef was exorcised, I hope forever. And with the help of Julian Muller, my editor at Harcourt Brace Jovanovich, I had even managed to clear a roadblock that had frustrated me for several years; my only work of *non*-science fiction, *Glide Path*, was finally ready for the printers. While I was wondering what to do next, Fate, disguised as Time Incorporated, settled the matter for me. I was offered an attractive fee, plus living expenses in New York, to write the text for the *Life* Science Library's *Man and Space*. (The title was my choice; I had completely forgotten it had been used only two years before by Ralph Lapp. I can only blame that blow on my head, and hereby apologise abjectly to Dr. Lapp.)

So during the early months of 1964, sitting quietly in Colombo, I wrote the main text of *Man and Space* and prepared for my encounter with the *Life* Book Division's gimlet-eyed editors (and researchers). The resulting volume, updated several times, has now sold over a million copies, in many languages. It would have been nice to have a share of the loot, but, alas, royalties are hardly practical for works of this nature, which involve the labor of literally dozens of artists, scientific advisors, writers—and *re*writers.

I would like to record, here and now, that, despite all the horror stories I had heard about working at Time-Life, I look back on this period with nothing but pleasure, and always make a point of revisiting my old friends in Rockefeller Center when I am passing through New York. Of course, I was a pampered visitor, not a regular staff member subject to deadlines and office politics; doubtless that makes quite a difference.

Man and Space, though a challenging assignment, was thus a pleasurable one and, as it turned out, a stepping stone to something enormously more important. But when I had finished the first draft a kind of intellectual backlash occurred, and I felt the need to relax.

Someone once defined a crank as an enthusiast without a sense of humor, and I have always believed that nothing is so important that you cannot make fun of it. (Perhaps I am the only person to joke about the end of the world; see "No Morning After" in *The Other Side of the Sky*.) At the same time, I must admit that I have occasionally been accused of having a schoolboy sense of humor. So I had better warn you that if you don't like the next item— there is even worse to come.

First of all, there are *two* kinds of space, and they mustn't be confused.

The old-fashioned variety, now badly dated, was discovered by a Greek named Euclid around 300 B.C. Not only did he discover it; he found out more about it than any sensible person would want to know. Not until two thousand years later did one Nikolai Ivanovich Lobachevski (Russian, though you'd never guess it) find that Euclid had cheated by assuming half the things he'd pretended to prove. But by then the damage had been done.

Perhaps we'd better distinguish between the two types of space by spelling one of them with a capital. Small s space—the variety that Euclid got hung up on—is what stops everything from being in the same place. This definition has the simplicity of genius; I thought of it only five minutes ago.

So much for space; now about Space.

The biggest difference between Space and space is that there is so much more of the former. Yet, strangely enough, it is much more expensive. Unless you live in New York, space costs practically nothing. But even a smidgen of Space costs millions, which is why only rich and powerful nations like the United States and the Soviet Union can afford to buy it. This is very annoying for ex-r. and p. nations like Britain, who often go round telling everyone who'll listen that Space isn't worth it. Another Greek, Aesop, had a term for this: he called it sour grapes.

Until about three hundred years ago, nobody realised that there was so much Space, and the philosophers were very unhappy when Galileo showed it to them through his telescope. Some people tried to pretend it wasn't there, but this did not work. Ever since men started looking at Space, it has grown bigger and bigger. When last measured, it took twenty-two zeros to express its size, and you can bet that it's grown some more since then.

For a couple of hundred years, no one did much about Space except look at it. A few crazy writers—if that is not a tautology—wrote stories about *going* into Space. Nobody took them seriously, which was just as well because most of their Space stories were excuses to poke fun at the existing state of affairs. If the authors hadn't set their adventures in imaginary planets, they would have gone to very real jails. (However, it is always dangerous to send authors to jail. This removes their chief excuse for not writing.)

At the beginning of the twentieth century there was a big change in the approach to Space. A Russian school-teacher named Konstantin Tsiolkovsky started to write serious scientific papers about travelling into it. Presently he was followed by Robert Goddard, an American college professor, and Oberth, a Rumanian high-school teacher. You will notice that all three pioneers of space travel were teachers: they were clearly trying to get away from *something*, and maybe it was the same in each case. All these early studies of Space pointed to one answer— the rocket. The rocket had been known for about a thousand years, having been invented (like everything else) by the Chinese, though they didn't get round to it until A.D. 1200, which was rather late in the day for them. All this time, however, the rocket had been used only as a firework and a weapon. (Incidentally, there has been an unfortunate misunderstanding over the phrase "The rocket's red glare." It is widely believed in the United States that these were *British* rockets, but a careful search of British history books gives no evidence to support this ridiculous suggestion. It must have been another firm of the same name.)

The world's hard-headed, practical engineers were so busy copying their grandfathers' mistakes that it was about fifty years before they realised what Tsiolkovsky and Company were saying, and as usual it took a war to put across the message. However, when the German V.2 started climbing to the edge of the atmosphere, it became rather difficult to ignore Space, though some people—as in Galileo's time—went right on trying.

But not in Russia, where Tsiolkovsky was a national hero. The U.S.S.R. went full steam ahead to develop long-range rockets, using the results of the German war effort and as many of the engineers involved in it as could be rounded up. Unfortunately (or not, depending on the point of view), it couldn't get the top rocket experts. Almost without exception, they had been shipped to the United States, where they sat chewing their fingernails for five years while leading American scientists explained that intercontinental ballistic missiles were science fiction. The Russians happen to like science fiction, so they launched the first ICBM in 1957, and the first satellite a few months later. Though they get most upset when told that their Sputnik launchers were built by German scientists, this is entirely their own fault. They won't release the names of their own top men, though everybody except the C.I.A. knows that they are called Korolov,[1] Glushko, Tikhonravov, Kostikov, Dushkin, and Robedonostev.

The discovery that the Russian satellites were about a hundred times as big as the American ones was, of course, a terrible blow to American pride, and vigorous steps were taken immediately to put matters right. These included setting up forty separate congressional committees on science and astronautics, so that all the top rocket scientists in the United States had to spend more than half their time giving evidence, preparing to give evidence, or recovering from giving evidence. Several million dollars were also appropriated for publicising such reassuring statements as "We are not engaged in a basketball game in space" and "Well, it's just a hunk of iron" every time a new Sputnik went up.

It was seven years before the United States was able to launch a satellite bigger than the Russians'—and when they did do so, most of the payload was five tons of Florida sand.[2] Later rockets will orbit used razor blades, thus solving one of the major problems of modern civilisation.

This leads us at once to a question which is always of some interest to the taxpayer—the *practical* value of

---

1. Sergei Korolov's name emerged two years later, in 1966, when he was given a state funeral and revealed to be the hitherto anonymous "chief designer" of the Soviet space program. A (presumably) highly fictionalised movie of his life, *The Taming of Fire*, appeared in the early seventies. Technically superb, especially in its historical reconstructions, perhaps the most astonishing thing about this film is the often unsympathetic, though in many ways admirable, character of the hero (he even fathers an illegitimate child). I don't know how the director got away with it, and perhaps he didn't.
2. Actually 11,500 pounds (Saturn I SA-5, 29 January 1964).

space research. Already it has produced thousands of valuable devices and discoveries which will soon make life easier for everyone. (These byproducts used to be referred to as "fall-out," but as that made everyone think of dandruff and H-bombs, the accepted term is now "spin-off." It is very popular with Air Force generals and top NASA brass, especially around budget time.) Here are just a few examples of spin-off:

Inertial guidance systems: These were developed to steer ICBM's to their targets, and have now reached a fantastic degree of accuracy. With an inertial guidance system in your car, you need never be lost, even in Brooklyn. You merely set it to zero at your starting point, and thereafter it tells you exactly where you are. This may not be the place you actually want to get to, but, by heaven, you will *know* where it is, and all for less than one hundred thousand dollars. Of course, if you don't know where you've started from in the first place, that's too bad.

Computers: The so-called electronic brains are certainly the most spectacular, though not the noisiest, products of the Space Age. They can do the most elaborate calculations in a fraction of a second, and it's hard to imagine how we ever got along without them in the past. Every one of us, at some time or other, has known the acute embarrassment of being unable to remember the square root of π, to solve a ninth order nonlinear partial differential equation with complex coefficients, and similar household chores. Those days are now gone forever.[3]

Miniaturisation: Some electronic components are now so small that more time is spent looking for them than using them. The transistor radio, which has been such a blessing

---

3. They certainly are. Who could have dreamed, back in 1964, that just ten years later millions of housewives would be able to tell you, in about two seconds, that root π is 1.772453851— and that slide rules and mathematical tables would be utterly obsolete?

to all lovers of peace and quiet, will soon be no larger than a pinhead. This will make tuning a bit tricky, but you can't have everything.

Vacuum techniques: Since Space is practically empty, this means that enormous quantities of vacuum will soon become commercially available; in fact, the import of vacuum by space freighters will be one of the first astronautical industries. Vacuums, as everyone knows, have thousands of uses, from sucking up dust to keeping drinks warm (or cool) on picnics. They also provide the best possible soundproofing, and so will allow apartments to be built with walls only a fraction of a centimeter thick, instead of almost a whole centimeter as at present. This will allow an extra quarter of a million people to live on Manhattan, and will increase the value of real estate everywhere. Portable vacuum sound-screens (probably collapsible) will be the only answer to the pin-sized transistor radios mentioned above.

Communications satellites: When these are perfected, the average man will have a choice of not less than ten thousand TV channels. Studies made by a leading Madison Avenue agency have already revealed some most interesting facts about the programs they are likely to carry. For example, of the ten thousand channels, not less than six thousand will be devoted to Westerns, and about three thousand will deal with crime. Since these programs will be broadcast over the whole world, travellers will no longer have to miss their favourite entertainment when they go abroad. Thus visitors to Switzerland or Bali won't have to spend hours looking at boring scenery; their portable TV sets will keep them in touch with the real world.

It has been calculated that, by 1985, every adult American will appear on a panel show at least once a month, and because of the enormously increased importance of ratings and surveys, a whole new professional class will spring up—the full-time TV viewer. This will be a great boost to the economy, as it will absorb thousands of citizens who, owing to their low IQ's, would be otherwise unemployable.

By the way, at least three of the ten thousand TV channels mentioned above will be largely devoted (apart from commercials) to educational and cultural matters.

Explosive-forming: This is a way of manufacturing quite complicated objects *instantly* by detonating suitably arranged charges of explosives. When this technique is perfected, you will be able to buy a prepackaged house which you can set down anywhere, light the fuse, and then walk in. When you're tired of it, you'll only have to buy some more explosives.

These are some of the down-to-earth applications of Space travel, but of course its most important result will be to take us to the other planets. And these are perfectly fascinating places.

Consider Mercury; when you go there, you'd better wear lightweight tropical clothing, as the thermometer can reach 750 degrees in the shade. And it's no good waiting for the cool evening breeze, because (a) there's no breeze and (b) there's no evening. The sun always stays put in the same place, apparently stuck in the sky.[4] The only way to cool off is to go to the night side of the planet; and then you will cool off nicely, to about 400 degrees below zero. This is what the tourist leaflets call "bracing."

Venus, at the moment, is something of a mystery. (Actually, she's been a mystery for three hundred years, and the more we learn the more the mystery deepens.) Though she is farther from the sun than Mercury, she is almost as hot, owing to an atmosphere a hundred times as dense as Earth's. And she rotates *backwards* on her axis, which is something no other planet would dream of doing.

---

4. A few years later it was discovered that Mercury does turn, slowly, beneath the sun. The improvement in the weather is negligible.

The fact that she is the only female in the solar system may explain this contrary behaviour.

Then, of course, there is Mars. Unfortunately, Mars is full of Martians. Though there is considerable disagreement among writers on the subject, almost all are agreed that you wouldn't want a Martian to marry your sister. Even when they look remotely human, they usually have unflattering skin tones. (Green is a favourite colour.) Some have even more deplorable characteristics, like mouths in the palms of their hands, or gigantic tusks (see Edgar Rice Burroughs). Even worse were the Martians discovered by H. G. Wells, who looked rather like squashed octopods and were addicted to human blood. Against the prejudice generated by all this propaganda, the writings of the "let's be friendly to the Martians" school, though headed by such able advocates as C. S. Lewis, Ray Bradbury, and Robert Heinlein, have so far made very little headway.

Mars also presents a number of health hazards to visitors. Among the indigenous diseases are Cimmerian Swamp Rot, a fungus that grows with explosive violence and can turn a man into something resembling a hairy doormat in ten minutes flat. Mention should also be made of the Xanthean Brain Leech which burrows into the spinal cord and takes complete mental possession of its victim. Even more dangerous is a type of hypnotic squid found in the lowlands, whose procedure makes the fabled sirens of Ulysses look like very small-time operators. By projecting irresistibly attractive visions into the mind of its prospective meal, it can lure any intelligent creature within range of its tentacles.

Life on Mars is tough, and one must be prepared for such little surprises.

The giant planets beyond Mars have fewer tourist attractions. Jupiter, the largest, is not the place to go if you are overweight; a terrestrial hundred kilos would turn into a quarter of a ton on Jupiter. The atmospheres are also slightly poisonous (hydrogen, methane, and—

phew—ammonia) and are so dense that you could skin-dive in them. However, the views from the nearer moons are magnificent, especially in the case of Saturn.

It seems unlikely that the outer planets harbour any natives, friendly or otherwise. As in the case of Mars, however, writers on the subject have assumed the worst, and have created a whole legion of Bug-Eyed Monsters, familiarly known as BEM's. Whatever their shape (if, indeed, it remains fixed for more than a few minutes at a time), all BEM's have two characteristics in common. They are implacably hostile; and they have an extraordinary and unquenchable interest in the female of the human species. Just what a BEM would do with one if it caught her has never been satisfactorily explained; in the interests of science, I cannot help wishing that someday a BEM would not be interrupted before it had finished whatever it wanted to do. A Society for the Prevention of Cruelty to BEM's seems well overdue.

I'd like to tell you a lot more about Space, but I'll have to break off here as something odd seems to be happening outside my window. There's a bright light flashing from a most peculiar object, hovering a couple of metres above the lawn, and I'd better investigate.

I can't imagine why such a ridiculous analogy popped into my mind—but it looks just like a saucer.

7

A Breath of Fresh Vacuum

Early in 1964, just as I was preparing to leave Ceylon to work on *Man and Space*, lightning struck in the form of a short letter from Stanley Kubrick. He said he was interested in doing a science-fiction movie and suggested we get together when I arrived in New York.

This was stimulating, but I refused to take it too seriously, knowing the mortality rate of movie projects. (Almost ten years earlier, I had sold *Childhood's End* to a Hollywood producer. Though *Variety* announces its imminent production about once a year, that's the last I've ever heard of it.) However, I looked forward to meeting Stanley, since I had admired *Lolita* and had heard excellent reports of his latest movie, *Dr. Strangelove*.

So now I had two objectives in New York—one a firm commitment, the other completely pie-in-the-sky (or beyond). Unfortunately, leaving Ceylon turned out to be exasperatingly difficult.

It had always been a matter of considerable annoyance to me that my friends assumed I lived in Ceylon because it was a tax haven—a sort of oriental Switzerland or Liechtenstein. In fact, the exact reverse was the case; the local rate of tax was very high. (One budget, fortunately short-lived, proposed a tax-rate of *more* than 100 percent.) So each time I left the country, I was involved in a protracted and financially debilitating encounter with the Inland Revenue Department.

On this occasion, it was exceptionally agonising; each time my accountants thought everything was settled, some new liability was discovered. When I finally left, I was rather like a Distressed British Seaman being shipped home by the local consul, with a fiver in his pocket. It was just as well for my peace of mind that I did not know that my remaining funds—the modest balance in my Manhattan bank—were about to be sequestered by an official whose existence would seem unlikely in this day and age. The "Sheriff of New York" sounds as if he should be riding down Wall Street, blasting away with a six-shooter in each hand. Wherever he hangs out, my wife's attorneys had located him, to my considerable embarrassment. Fortunately, I had good legal advice (Cy Rembar, who later took on the entire Supreme Court in defence of literary liberty), and matters were sorted out amicably enough. The last time I spoke to my ex—let's see, that would be around 1970—we were on the best of terms, and I was quite touched to discover that she still used her married name. Hi, Marilyn!—wherever you are.

I mention these sordid but all too important details because for the next ten years the whole pattern of my life was to be controlled by lawyers and accountants—a common fate of writers. Not until 1975, as I shall happily explain in due course, was I able to stop anxiously counting days and putting big red X's on calendars, to mark the deadlines when I had to cross frontiers in peril of instant bankruptcy.

As is now fairly well known, my meeting with Stanley Kubrick was quite successful, resulting in some 141 minutes of celluloid, a couple of books, millions of words of controversy, and greatly increased posthumous fame for Richard Strauss. Since the story of our collaboration has been described in *The Lost Worlds of 2001* (New York: New American Library, 1972) there is no need to go into details here; suffice to say that as Time-Life's *Man and Space* phased out in early 1964, *Journey Beyond the Stars* phased in.

Journey Beyond the Stars? That was the original title of the project, when Stanley announced it to the press. I never liked it, because there had already been so many indistinguishable (and undistinguished) science-fiction journeys and voyages. (Indeed, the inner-space epic *Fantastic Voyage*, featuring Raquel Welch and a supporting cast of ten thousand blood corpuscles, was also going into production about this time.) Other titles which we tinkered with but never took very seriously were *How the Solar System Was Won, Universe, Tunnel to the Stars,* and *Planetfall.*

As I was engaged in writing the novel and screenplay in the United States—specifically, in Room 1008 of the famous Hotel Chelsea on New York's Twenty-third Street—it was necessary for me to become a United States resident, and to go through all the formalities of obtaining a Resident Alien card. This document always made me feel like a certified extraterrestrial, which seemed highly appropriate under the circumstances.

However, it did complicate my image and make it hard to define my status; it was easy to sympathise with the bafflement of journalists when I explained that I was a British citizen, an American resident, and a Ceylon householder. In fact, I was to be exiled from Ceylon for well over a year, and it is not surprising that during the almost three years of scripting and shooting (at the M.G.M. Elstree Studios, just north of London) I had no time for any writing that did not directly concern *2001.*

This included a lot of publicity material, duly sent out in all directions by M.G.M. Though it contains some of my finest purple prose, I do not think that it is worth reproducing here, and I rather suspect that *2001* would still have been quite a success without it.

However, there is one short essay I *would* like to preserve, as it was written to answer criticism which still appears from time to time. Perhaps the most dramatic scene in *2001* is that in which David Bowman, after HAL's mutiny, reenters *Discovery* without benefit of space helmet, thus exposing himself to vacuum for about ten seconds. (After all the screenings I've sat through, I *still* hold my breath for the duration of this sequence.)

This, I knew, would cause protests, so I decided to attack first. Although "A Breath of Fresh Vacuum" was written in 1966, nothing has since emerged to change my conclusion.

For some years I have been trying, in a rather desultory fashion, to get the space medics interested in the possibility of man's survival in vacuum. To the best of my knowledge, it was the brilliant and, alas, short-lived, young science-fiction writer Stanley G. Weinbaum who first suggested that exposure to space need not be instantly fatal. In "The Red Peri" (*Astounding Stories,* November 1935) he gave plausible reasons for thinking that spacesuits might be unnecessary for short periods. His arguments carried conviction, though he probably took the idea too far. Yet in a recent story (Pierre Boulle's *The Garden on the Moon*) we have a character literally exploding when he tears his spacesuit; this is certainly nonsense.

What *would* kill a man on exposure to vacuum? Ultimately, of course, lack of oxygen; but this takes a considerable time, as any skin-diver knows. (Starting in my sedentary thirties, I was eventually able to stay down for over three and a half minutes; the record *without breathing gear* is almost a quarter of an hour!) It is obvious that, on this score alone, a man in good health should be able to go without external oxygen for at least a minute; and even ten seconds might make all the difference between life and death. In an emergency, you can do a lot of things in ten seconds.

The pressure change is a more immediate hazard; at sea level, the differential over the whole body is about fifteen tons, and from the lowest pressure at which men could comfortably breathe pure oxygen (say a fifth of an atmosphere) the change when going to complete vacuum would still be about three tons. Hence the gory visions of exploding astronauts.

But, of course, the human body is mostly incompressible liquid, and only its internal air spaces are of importance here. Not long ago, doctors were denying that men could skin-dive to depths of thirty metres; the record is now seventy,[1] a change of six atmospheres, up and down, in two minutes. As long as the main air cavities could be kept open, so that there was no pressure build-up, exposure to vacuum should not cause damage.

But what about the boiling of the body fluids? There have been impressive photos of astronauts in vacuum chambers, holding beakers of water which turn instantly to vapor. The thought of this happening to one's blood is enough to discourage the most intrepid experimenter. However, once again the human body is a very tough proposition; I am indebted to Dr. Gene Konecci for the striking phrase, "The skin makes a pretty good spacesuit." Moreover, the word "boiling" in this context is very misleading. Vaporisation occurs at body temperature, and because the latent heat necessary has to come from somewhere, the net result is actually freezing rather than boiling!

I first worked out these ideas in the novel *Earthlight* (1954) and in more detail in the short story "Take a Deep Breath" (1957). Now my colleague Stanley Kubrick, in the forthcoming movie of our joint novel, *2001: A Space Odyssey,* has shown the whole process. (In Cinerama, yet.) This will present the concept of vacuum survival to millions who are quite certain that it won't work . . . and thousands of "experts" who would swear to its impossibility on a stack of Bibles.

---

1. By 1976 it was well over a hundred metres!

I am very happy, therefore, to see the results from Holloman and Brooks AFB's on the decompression of chimpanzees and dogs to near vacuum (less than 2 mm) which completely support my thesis. The chimps survived for periods of up to two and a half minutes; it is hard to believe that the residual 1/400th of an atmosphere could have made a significant difference to their chances.[2] The dogs did about equally well; all survived up to two minutes, but mortality increased progressively from two to three minutes. (The researchers apparently regard dogs as expendable; *all* the chimps recovered, so were clearly taken nowhere near their limits.)

If animals can survive this treatment, then it is obvious that a trained man, *who knows what to expect and can prepare for it,* for example, by exhaling and keeping his mouth open before decompression, would do much better. Until the experiment is attempted we won't know for sure; it's the sort of thing the late, great J. B. S. Haldane would love to have tried. Any volunteers?

For the moment, it certainly seems reasonable to assume that men can expect at least ten seconds—perhaps much more—of useful consciousness in vacuum. And even when unconsciousness ensues, if they are recompressed within a minute or two they would suffer no permanent ill-effects.

Having gone thus far, I will stick my neck out farther. I would not be in the least surprised to find that vacuum exposure is just another of those space bogeys, like meteorites and weightlessness. One day, lunar colonists in a hurry may hop from vehicles to air locks, over distances of a few yards, without bothering to suit up—just as

2. *The Effect on the Chimpanzee of Rapid Decompression to a Near Vacuum.* Edited by Alfred G. Koestler, NASA CR-329 (November 1965). Also *Experimental Animal Decompressions to a Near Vacuum Environment.* R. W. Bancroft & J. E. Dunn. Report No. SAM-TR-65-48 (June 1965), USAF School of Aerospace Medicine, Brooks AFB, Texas.

Cousteau's inner space explorers wander around three hundred feet down *without* aqualungs when the mood takes them.

And if anyone shouts "Rubbish!" I invite him to step outside and say it again.

8

The World of 2001

It is hard to realise that 1984, with all its ominous resonances, is just around the corner—and thereafter it's a straight run of only seventeen years to 2001. Although the movie had (very wisely) completely bypassed the terrestrial scene and shown nothing of everyday life at the beginning of the next century, I was constantly being asked to lower my sights from the moon and beyond, and to write factual forecasts, rather than science fiction.

This was something I was reluctant to do, for a whole constellation of reasons. As the record of the past shows, the most interesting and important events are usually those that were never predicted; history, like much of physics, is often discontinuous. (A recent example is the Arab oil embargo. Of course, that traumatic event was easy to predict—*afterwards.*)

Moreover, I have always been careful to point out that science-fiction writers very seldom attempt prophecy; they are interested in *extrapolation,* which is a very different matter. The S.F. writer is always asking himself "What if . . ." or "Just suppose. . . ." He very seldom says "This is the way it *must* be." What he may say—which is exactly what Orwell was trying to do—is "This is the way it will be . . . *unless.* . . ."

Ray Bradbury once summed up the situation beautifully. "I don't try to predict the future," he remarked. "I try to *prevent* it." (Read *Fahrenheit 451* to see what he meant.)

Nevertheless, some elements of what may be called the technological future are obviously predictable, at least in the short term, and with a limited degree of accuracy. "Futuristics" is now a vast industry, or possibly a religion, with Herman Kahn as its patron saint. (Except that Herman has too much of a sense of humor to make a very plausible saint; and it is rather hard to imagine him fasting.)

So in 1966 I took time off from the movie, at the request of *Vogue* magazine, to give my ideas about the world we might see in 2001. This article later formed the basis of a lecture I have given all around the world, but it has never appeared between hard covers in its pristine form. I have deliberately refrained from altering it in any way—even to the extent of the opening chronology—because it is unfair and misleading to update articles about the future; they should be left in their correct time context, so that one can judge the truth of a slogan I once framed: "The future isn't what it used to be."

Vogue must be read in Moscow, for to my surprise the article was reprinted by that scourge of decadent capitalist culture, the *Literaturnaya Gazetta*. To my even greater surprise, I was duly paid in U.S. dollars, though I had been plaintively telling my Russian publishers for years that Ceylon rupees were entirely acceptable....

This was the very first time I got any money out of the U.S.S.R. But it was not, I am happy to report, the last.

———

On 1 January 2001, the second millennium ends and the third begins. It will be a time of stocktaking and good resolutions and, doubtless, of thankful farewells to the unlamented twentieth century. What kind of world will salute the first day of the new era when it dawns thirty-four years from now?

Any attempt to predict the future can be based only on experience of the past. The year 2001 is now as remote from us as the early 1930's—and *they* already seem to belong to another world. Yet there are reasons for expecting far greater changes in the three decades to come than in those that have just passed, for the pace of technology doubles every ten years or so. On this basis, 2001 may be as strange to us as our age would be to a man from 1890. In some fields—for example, science and space exploration—this will certainly be true; whether it will be true of ordinary, everyday life is a different matter. For it is possible that most of the profound changes to civilised man's mode of living have already taken place, and that we may expect no really revolutionary developments in the future.

This thought occurred to me rather vividly at the New York World's Fair, when I was one of the several million visitors to the General Electric exhibit. Here the ubiquitous Walt Disney had devised a rotating theatre to show how the home has changed since the Gay Nineties; the audience moved, as if in a time machine, in jumps of some twenty years up to the present. We watched gas lamps, coal-burning stoves, and hand-operated vacuum cleaners give way to air conditioners, large-screen TV, and the all-electric, automatic kitchen.

Upon one visitor, at least, this fascinating exhibit did not produce quite the impression that G.E. had intended. (Perhaps this may be because I live in the Orient, where to 90 percent of the population Walt Disney's 1920 kitchen would be a vision of an impossibly utopian future.) At any rate, it seemed to me that there had been little fundamental progress in the last forty years—only refinement of details. Before 1920, as far as the prosperous Western nations were concerned, the revolution had already occurred, and a pattern of domestic existence that had lasted with no change for thousands of years had become obsolete.

One can list, quite briefly, the main ingredients of that revolution—one so great that, from our point of view, a medieval palace now appears little better than a mud hut. Here they are, not necessarily in order of importance:

piped water; indoor plumbing; gas cooking and heating; electric light and power. All these came into general use around the end of the nineteenth century; in the thousand years that went before, I can think of only one comparable advance—the introduction of glass windows. (Anyone who doubts their importance should try to imagine a typical winter night at Camelot.)

If indeed the major technical upheavals of the home are already behind us, the future will show little more than changes of styling and location. There will doubtless be Louis XIV air conditioners, Victorian TV sets, Ming dynasty toasters, Cretan tableware, Pre-Raphaelite computer consoles. Men will build homes in ever-stranger places, as during the last decades they have invaded the once inhospitable desert. There will be luxurious dwellings on the slopes of Everest, in the Antarctic wastes, amid the coral paradise of the Great Barrier Reef. But most of the gadgets in those homes would be understandable to us—as most of *ours* would certainly not be comprehensible to a man of the eighteenth century.

This is the conservative view of the future; it may even be correct. But before we accept it, let us remember all those prophets in the past who were so sure that the future could be no more than a projection of their own age. We may laugh at their failures of imagination, but ours may be equally great. Perhaps, after all, the most astonishing changes may still lie ahead of us.

What could they possibly be?

Some of the most important will be changes in *function*, caused by economic and technical developments outside the home itself. There was a time when every family unit had to be virtually self-contained—making its own clothes, preparing its own meals, building its own furniture, doing its own laundry, baking its own bread, manufacturing its own soap (if any). Life a few centuries ago was really very complicated, in ways we have completely forgotten. A visit to Mount Vernon, with its rooms full of obsolete and now often incomprehensible tools and gadgets, provides a good reminder of this. Here, armies of servants toiled at myriads of specialised jobs to sustain the Washington family. The first President of the United States would have been quite surprised to learn that the time would come when none of his successors' households would contain a single hand loom or spinning wheel.

These things have been swept into limbo by mechanisation and mass production. Today, about the last manufacturing process left in the home is the preparation of meals, but the time is fast approaching when home cooking will be as rare as home weaving. The freezer and the TV dinner together have combined to doom the kitchen, in which our mothers toiled away most of their lives.

A whole month's meals for a family could be delivered, frozen and/or dehydrated, in a package weighing about fifty kilograms. (Remember that 90 percent of most food is water; why carry that home when you can get it from the tap?) The unit would be placed in a kind of "home automat," the desired meal would be selected, and the built-in computer would do all the rest. A perfectly cooked meal would pop out of the dispenser in less than ten minutes—or a sign would flash saying "Sorry, filet mignon out of stock."

And if it *was* in stock, it would probably never have been near a cow. The production of natural meat is so inefficient a process that it may be uneconomical, or even prohibited, by the twenty-first century. If the present population increase continues, for every man who consumes a meal based on meat, ten other men will have to starve.

The logic of this is inexorable, for it takes roughly ten kilograms of vegetable matter to make a kilogram of meat, via cows, sheep, and hogs. It's true that these mobile food factories consume raw materials, like grass, which we cannot use directly. But the biochemists are making great progress; our grandchildren will *love* grass, and won't even know that they're eating it. It will taste like anything they care to specify—and, equally important, its appearance will match its taste.

Beyond the production of all types of food from plant or vegetable sources, we will undoubtedly see its manufacture from nonliving materials, such as coal and oil (too valuable to burn in the future, but not too valuable to eat), limestone, carbon dioxide, and water. Ultimately, purely synthetic food may be the only way of feeding our planet's billions; it is certainly one of the requirements for the colonisation of such barren worlds as the moon, which possess no life forms of their own that the explorers can eat. (It is to be hoped they also possess no life forms that will eat the explorers.)

The conquest of Space, which has forced the development of so many new technologies and materials, is going to change the pattern of everyday existence in innumerable ways. There has already been much talk of products coming into commercial or domestic use as a result of spin-off from the space program, but most of the examples quoted so far have been rather trivial, and certainly do not justify their development costs. I find it hard to generate much enthusiasm for unbreakable, heatproof kitchenware made from the ceramics used in missile nose-cones. Most of us, I am sure, would have been happy to do without these ceramics, if we could also dispense with the missiles.

However, in the decades to come our homes may be completely transformed—not merely improved—by some of the products of space technology. Perhaps the most extraordinary possibility is one suggested by Professor Buckminster Fuller; it might be called the autonomous or self-contained house, and would virtually be a ground-based spaceship. For long-range journeys, lasting months or even years, spaceships will have to be completely self-sufficient, processing all waste products and converting them back into food, air, and water in a closed-cycle economy. The equipment to do this is now being developed, at a cost of billions of dollars; but once it is perfected, mass production could make it available at a price every house owner could afford.

And then we would have a very interesting situation, involving social changes at least as great as those produced by the automobile. Today's vast food production and distribution industries would virtually cease to exist, since every home could feed itself automatically and indefinitely.

It must be admitted that there is a psychological problem here, as well as an engineering one. The idea of consuming one's own wastes over and over again in an eternal cycle is not particularly attractive, but this is *exactly* what we have been doing ever since agriculture was invented some twenty thousand years ago. To primitive people who obtained their food by hunting, farming must have seemed a disgusting process, but today it appears completely natural to us. Indeed, there is now a vigorous body of opinion agitating against the use of chemical fertilizers, and insisting that we should use only what is euphemistically called "organic" ones! So psychological objections to the self-contained food-production system are quite illogical, and are not likely to prevail—particularly if the economic and social advantages are obvious.

Among the latter might well be the development of completely mobile homes. Given a compact power source (and we may expect this as a result of fuel-cell or atomic energy developments), the house of the future would have no roots tying it to the ground. Gone would be water pipes, drains, power lines; the autonomous home could therefore move, or be moved, to anywhere on earth at the owner's whim. The freedom that the automobile gave to our fathers may thus be only a foretaste of the mobility which our grandchildren may enjoy.

Since today's homes are still built with Stone Age materials, they weigh at least ten times as much as they should. The space laboratories have now produced wonderful new materials far stronger than steel, but lighter than aluminum; when these are generally available, houses will be able to fly. They could be carried from place to place by "sky cranes" no more powerful than some of the big helicopter-freighters in use today. The time may come, therefore, when whole communities may migrate south in the winter, or move to new lands whenever they feel the need for a change of scenery. Anyone who thinks that there is no room left for such migrations has not looked very carefully at the terrestrial globe. There are vast and completely uninhabited areas of this planet—some of them extremely picturesque—which are empty simply because they are useless for food production. But these are just the regions that will be attractive *when* we have developed the autonomous home. The swimming pool in the Arizona desert is, perhaps, a symbol of the future in more ways than one.

It is clear that a mobile, planetwide culture demands a cheap, instantaneous, and universal system of communications. This is already on the horizon, and it is impossible to underestimate the impact it may have upon business, culture, and indeed every aspect of human life. The changes wrought by the telephone during the last ninety years may be negligible in comparison. For the telephone is an inflexible instrument, bound to fixed points by tendrils of wire, and capable of handling only one type of very slow-speed message—the spoken word. Compared to some of the devices now available, it is barely one step ahead of Indian smoke signals.

The approaching communications explosion has been detonated by a whole series of novelties: transistors, satellites, lasers, and other inventions whose names will probably never be heard by the general public. Individually, they were important breakthroughs, but their *combined* effect will be far larger than their sum. It will, indeed, be revolutionary—perhaps even catastrophic.

Without going into details, let us list just a few of the services which will be available by 2001—not in the office or factory, but in *every home:*

Direct TV reception, via satellite, from all major countries and political groups. This will imply several hundred channels, some of them of an extremely high cultural level.

Cheap telephone calls between all locations and even *moving individuals* everywhere on earth. They will be billed at a flat rate, or possibly not even billed at all, if the equipment is hired.

On a slightly more expensive basis, worldwide face-to-face TV conversations.

Facsimile services whereby letters, printed matter, etc. can be reproduced instantly. The physical delivery of mail and newspapers will thus be largely replaced by the orbital post office, and the orbital newspaper, beaming down signals from space.

Immediate access in the home via simple computer-type keyboards, and TV displays, to all the world's great libraries and information centers. Any items needed for permanent reference could be printed off as soon as located on a copying machine—or filed magnetically in the home storage system.

A computer to supervise and perhaps control all the normal household chores: hygienic, secretarial, culinary. A kind of benevolent central intelligence, it would combine most of the duties of butler, accountant, and social secretary. It would probably be called Jeeves.

This will do for a beginning, but it will already be obvious that with such electronic aids, a great many of today's social patterns will be changed out of recognition. (Remember what home life was like in the pre-TV era? The pre-*radio* era?) One of man's oldest and—occasionally—noblest, inventions will be rendered obsolete. The city will begin to die.

It is probably dying already, though of other causes. When men everywhere can meet at a touch of a button, far more cheaply and conveniently than finding a cab in a Manhattan rainstorm, they will choose the easier way of life.

Small cities, or large towns, may still be necessary for the conduct of industrial processes; and doubtless the traditions of such university towns as Oxford and Ann Arbor will continue even in the age of teaching machines and TV lectures. But the vast conurbations that have blighted so much of the world for two centuries or more will slowly fade away. *Very* slowly, because bricks and mortar and steel and concrete have enormous inertia and represent such a vast capital expenditure. I have little doubt that there will be even larger cities in the year 2001 than there are today, but they will be like the dinosaurs in the last age of their giantism. A century later, there will be only bones.

Unless . . . there is always a possibility that the population explosion cannot be controlled, and in that case the whole world will become one seething city. But this will not last for long; though medical science may keep that old regulator, the Black Death, at bay, Nature will somehow redress the balance. There are plagues of the mind worse than those of the body, and the meaningless violence that has already shattered many communities could be a mild foretaste of a psychologically overcrowded future. Though everyone now accepts the need for population control, very little thought has been given to the level that should be aimed at. This involves questions that are very difficult to answer, but which determine the entire structure of future society quite as much as do any technological changes.

There is no doubt that, with proper organisation, our planet could support a population of many billions at a much higher standard of living than today. But should it? In a world of instantaneous communication and swift transport, where all men are virtually neighbors, is there any point in a population of more than a few million? The answer to this question depends upon one's philosophical and religious views concerning the purpose of life. Fred Hoyle, for example, once suggested to me that the optimum population of the world should be about a hundred thousand, as that was the maximum number of people one could get to know in a lifetime. And it is worth remembering that Plato thought the ideal city should contain only about five thousand "free men." However, Plato's city also contained a much larger number of slaves; his so-called democracy could not have functioned without them. Nor can the world of the future, especially if its population is ultimately stabilised at a fraction of today's figure. The big difference is that these "slaves" will not be human.

Most of them will, of course, be robots at all levels of sophistication from naïve automatic washing machines to the ultraintelligent home computers already mentioned. The great majority will be fixed, though their sense organs and manipulators will be distributed over a wide area—like today's central heating systems, which may have thermostats all over the house. Some will be mobile, doing cleaning jobs and looking after the garden, but for this sort of task, there is a nonmechanical solution which offers not only economic advantages but an emotional bonus.

Why go to the trouble of building extremely complex mobile robots when Nature has already done 99 percent of the job for us? During the last decade, the ultimate secrets of the living world have been revealed; the mechanisms of the cell and of heredity have been uncovered. The next great breakthrough in technology will be *biological engineering:* the design of organic, though not necessarily living, systems to carry out any desired task.

Animals, of course, have long been used by men as extensions of their personalities or their bodies. The sheep dog, the working elephant and, more recently, the guide dog for the blind, are quite remarkable examples of what can be done with already existing animals and primitive training techniques. If we tried, in a few decades we could develop a creature based on the chimpanzee, but with a tenfold improvement in intelligence, motivation,

vocabulary—and disposition. Such a domesticated ape
(Pan sapiens?) might be produced by a combination of
genetic selection and biological engineering; when it
comes on to the labor market, the servant shortage will
be over—and the housewife of 2001 need no longer be
envious of her great-grandmother in 1901.

You may well object at this point—if you didn't a good
deal earlier—that the main result of all these developments will be to eliminate 99 percent of human activity,
and to leave our descendants faced with a future of
utter boredom, where the main problem in life is deciding
which of the several hundred TV channels to select. This
is perfectly true—*if* we look at humanity as it is con-
stituted today.

H. G. Wells once said that future history would be a
race between education and catastrophe. I doubt if even
that great prophet realised the educational standards that
would ultimately have to be attained to cope with the
problem of universal leisure. For while, ironically enough,
one of the slogans of today's politicians is "full employment," what we are really heading for is the exact reverse: full *un*employment. Just as there is little place
today for manual laborers, and there will be none tomorrow for those of only clerical and executive skills, so
in the day-after-tomorrow society there will be no place
for anyone as ignorant as the average mid-twentieth-
century college graduate. He would be as lost and helpless
as one of the Pilgrim Fathers, suddenly dumped in Times
Square during the rush hour.

If it seems an impossible goal to bring the whole population of the planet up to superuniversity levels, remember that a few centuries ago it would have seemed equally
unthinkable that everybody would be able to read. Today
we have to set our sights much higher, and it is not
unrealistic to do so. The development of teaching machines, current investigations into the way in which the
brain stores information, experiments with the so-called
consciousness-expanding drugs—these hint at the new and

revolutionary weapons our children may be using in the war against ignorance. It has been said that the greatest single industry of the future will be education. This may well be true, because for every man education should be a process which continues all his life. We have to abandon, as swiftly as possible, the idea that schooling is something restricted to youth. How can it be, in a world where half the things a man knows at twenty are no longer true at forty—and half the things he knows at forty hadn't been discovered when he was twenty?

So perhaps the greatest change, at the turn of the millennium, will be in the mental attitudes of our descendants, rather than in the physical backgrounds of their lives. They will have developed a flexibility of mind, a capacity for organising knowledge, and an active curiosity about the universe which will make them almost a new species. Anyone who has had the stimulating—and chastening—experience of meeting a *really* educated man will know exactly what I mean. (I've been lucky—I've met three.)

In the race against catastrophe, of which H. G. Wells warned us, the last lap has already begun. If we lose it, the world of 2001 will be much like ours, with its problems and evils and vices enlarged, perhaps beyond endurance. But if we win, 2001 could mark the great divide between barbarism and real civilisation. It is inspiring to realise that, with some luck and much hard work, we may live to see the final end of the Dark Ages.

9

"And Now—Live from the Moon..."

In 1968 the two events which filled my field of consciousness to the exclusion of almost all else were rushing to their climax on parallel tracks. Some time in the spring *2001: A Space Odyssey* would have its premiere. And towards the end of the year, if all went well, the first Apollo astronauts would leave for the moon.

There was nothing much more that I could do for *2001*, which Stanley Kubrick was now cutting round the clock, but I had managed to get involved in three separate and simultaneous projects relating to Apollo. The CBS network had asked me to join Walter Cronkite and Wally Schirra during their TV coverage of the missions. With the celebrated documentary producer Francis Thompson, I was scripting a wide-screen movie which M.G.M. and Time-Life were jointly financing; and I had agreed to write a twenty-thousand-word epilogue to the official volume which would be published when the astronauts returned.

The TV coverage was in itself the experience of a lifetime, with its trips to Cape Kennedy, the excitement of the launch, the long sessions in the New York studio waiting to go on the air at critical moments—the whole thing held together by the tireless and unflappable Walter Cronkite.[1] But apart from miles of video tape and a beautiful souvenir volume put out by CBS News (under

---

1. I have his word for it that a recent unlamented President called him "the best of a bad bunch." Woodward and Bernstein, were you listening?

the librarian-baffling title 10:56:20 PM EDT 7/20/69),
there is no other record of the hours I sat in front of the
cameras. Perhaps it is just as well, for the remarks made
spontaneously under the stress of emotion seldom read
well in print. I'm not sorry that the words I spoke in those
historic moments are now more than sixty trillion kilometres away from Earth, still heading outwards at three
hundred thousand kilometres a second.

I *did* contribute a number of short essays to the CBS
stockpile, for possible delivery during those long intervals
when nothing much was happening on or between Earth
and moon. As it turned out, they were never used, and
any qualms Eric Sevareid may have had about the com-
petition were unnecessary.

Here is one of these minilectures, which I gave what I
hoped would be an intriguing title: "Thirty Years of
Space Travel."

Thirty years ago, my life was dominated by the infant
British Interplanetary Society, of which I was treasurer
and general propagandist. We space-travel enthusiasts—
the hard core never consisted of more than a dozen people
—dreamed in 1939 that it would be possible to build a
rocket to go to the moon. Of course, hardly anybody believed us, and most people regarded us as complete crackpots.

We were in touch with similar groups in other countries,
especially the United States, where the American Rocket
Society was facing similar skepticism. We knew that Dr.
Robert Goddard had flown rockets to heights of a few
thousand feet, and we were aware that there had been
considerable activity in Germany. But we had no con-
tacts there—for reasons which became apparent later in
the war. German rocket research had become top secret.

Our attempts to interest the British government in

rocket propulsion resulted in a letter stating that scientific studies had shown that there was no practical use for jet propulsion. Well, we hadn't even been talking about jet propulsion. I wonder if anything ever came of *that*?

Since the society's annual income never exceeded three hundred dollars, the possibilities of actual experimentation were rather remote, perhaps luckily for us. So we concentrated on paper studies, and in 1938 published what were for the time quite detailed proposals for a spaceship to land on the moon. It would carry three people and descend on four shock-absorbing legs which would support it in the right position for a take-off and return to Earth. The whole design looks remarkably like that of the lunar module thirty years later, but we can't take too much credit for this. Similar engineering problems must necessarily evoke similar solutions.

After the war, in which we all learned a great deal about technology and, of course, saw the advent of the large liquid-fueled rocket, we continued our studies, and the *Journals* of the British Interplanetary Society of that time are full of papers about orbital rendezvous, space stations, and the use of specialised vehicles for such purposes as transfer from orbit to orbit, and for the actual landing on the moon. Here again we can't claim any great originality because all these ideas had been discussed by the German and Russian theoreticians decades before.

In the postwar years, when the V.2 had demonstrated that large rockets *could* travel beyond the atmosphere, our ideas were listened to much more sympathetically. Distinguished scientists were no longer ashamed to be seen in our company, and even joined the Interplanetary Society. But perhaps our most famous member at that time was a literary figure—none other than George Bernard Shaw, who joined the society in his ninety-first year and remained a member until his death. I am particularly proud of capturing him, as a result of a paper I sent him in 1947 on the philosophical aspects of space travel. This resulted in a correspondence about supersonic flight in which he put forward some very peculiar theories about aerodynamics.

Another literary figure who was not so sympathetic to our aims was the theologian and novelist C. S. Lewis. Although several of his best books are about space flight, he was very much opposed to the idea and attacked rocket societies because they would spread the crimes of mankind to other planets. This annoyed me, and we arranged a confrontation in a famous Oxford pub. My second was Val Cleaver, later head of the Rolls-Royce Rocket Division, and Dr. Lewis was supported by Professor J. R. R. Tolkien, since famous for *The Lord of the Rings*. We had a splendid time arguing about the merits of space travel, and as we parted Dr. Lewis said, "I am sure you are very wicked people, but how dull it would be if everybody was good."

Probably the most important thing we ever did was to arrange one of the first international congresses on space travel, in London during the summer of 1951. This congress was devoted entirely to the artificial satellite: it was four years later that the United States announced that such satellites would be launched. At this congress, experts described how one could build artificial satellites, and discussed all the useful things that could be done with them. All these predictions were to come true within the next ten years.

It's fascinating to look back on some of the ideas that were published in those early days. Almost all the *technical* details were dead right—practically everything that has happened in space travel was described ten, twenty, or even fifty years ago. But we were wrong in two important respects.

In the first place, we never dreamed of the incredible complexity and cost of spacecraft and launchers. That a space vehicle would have to contain millions of components, all of which must function perfectly, would have seemed to us impossible. If we had realised how complicated the actual engineering would be, and how much it would all cost, I am sure most of us would have been so discouraged that we'd have taken up something simpler, like building paper airplanes. In 1939 we thought that a spaceship to carry three men to the moon might be built for a million dollars. Well, the Apollo project cost over $20 billion. Even inflation doesn't quite account for the difference.

On the other hand, we greatly overestimated the time all this would take. Before the war, I doubt if anyone really thought that the moon would be reached in this century. When I published my first space novel in the early 1950's, I very optimistically imagined a lunar landing in 1978. I didn't *really* believe it would be done so soon, but I wanted to boost my morale by pretending that it might happen in my own lifetime.

Now I raise my sights much higher by hoping to see men land on Mars, and I annoy my conservative engineering friends by announcing my own intention of going to the moon one day, as a tourist. After all, if George Bernard Shaw joined the Interplanetary Society when he was ninety-one, why shouldn't I stay at the Lunar Hilton in my eighties?

———

As for the Apollo Cinerama spectacular, I have even less to show for that. Francis Thompson and I completed the script, after months of work, and a good deal of special material was shot, including a fascinating sequence showing Apollo 8 Commander Frank Borman's family, in their own home, watching TV at the moment when the words "They're on the way to the moon!" were spoken for the first time in history. But then M.G.M. ran into financial difficulties, the partnership with Time-Life was dissolved, and the whole project was cancelled. We were paid for our work, of course, but the epic-that-might-have-been never reached the movie screens of the world. Some years later, Francis Thompson did complete a more modest NASA-financed version, *Moonwalk One*, but it had only limited distribution, though it is probably the best record made of man's first contact with another world.

My twenty-thousand-word look beyond Apollo did, however, duly appear in *First on the Moon* (Boston: Little, Brown, 1970), and I had the privilege of sharing the title page with Neil Armstrong, Michael Collins, and Edwin Aldrin—none of whom, as it happened, I had met at the time. (I still haven't met "Buzz" Aldrin.)

I put my heart and soul into the epilogue of *First on the Moon*, and I am sorry that the book was not as great a success as everyone had anticipated; unfortunately, by the time it appeared, the United States was starting to slide into the time of troubles from which it has, one hopes, at last started to emerge, and interest in space travel was rapidly diminishing.

So I would like to quote two very short extracts from the ending of the book, because in a few words they summed up all that I had been trying to say in a lifetime.

Five hundred million years ago, the moon summoned life out of its first home, the sea, and led it onto the empty land. For as it drew the tides across the barren continents of primeval earth, their daily rhythm exposed to sun and air the creatures of the shallows. Most perished—but some adapted to the new and hostile environment. The conquest of the land had begun.

We shall never know when this happened, on the shores of what vanished sea. There were no eyes or cameras present to record so obscure, so inconspicuous an event. Now, the moon calls again—and this time life responds with a roar that shakes earth and sky.

When a Saturn V soars spaceward on nearly four thousand tons of thrust, it signifies more than a triumph of technology. It opens the next chapter of evolution.

No wonder that the drama of a launch engages our emotions so deeply. The rising rocket appeals to instincts older than reason; the gulf it bridges is not only that between world and world—but the deeper chasm between heart and brain.

Whether we shall be setting forth into a universe which is still unbearably empty, or one which is already full of life, is a riddle which the coming centuries will unfold. Those who described the first landing on the moon as man's greatest adventure are right; but how great that adventure will really be we may not know for a thousand years.

It is not merely an adventure of the body, but of the mind and spirit, and no one can say where it will end. We may discover that our place in the universe is humble indeed; we should not shrink from the knowledge, if it turns out that we are far nearer the apes than the angels.

Even if this is true, a future of infinite promise lies ahead. We may yet have a splendid and inspiring role to play, on a stage wider and more marvelous than ever dreamed of by any poet or dramatist of the past. For it may be that the old astrologers had the truth exactly reversed, when they believed that the stars controlled the destinies of men.

The time may come when men control the destinies of stars.

10

Time and the Times

Of my numerous pieces of journalism relating to the Apollo project, there are two I would like to preserve, because they help encapsulate a moment in history when —as never before, and when again?—all men's minds were turned to a single thought.

On Thursday 17 July 1969, the day after launch, the *New York Times* brought out a special forty-eight-page supplement devoted entirely to the Apollo 11 mission. My contribution appeared under the title (not mine), "Will Advent of Man Awaken a Sleeping Moon?"

The supplement also printed, with a straight face, this belated correction:

On January 13, 1920, "Topics of the Times," an editorial-page feature of the *New York Times,* dismissed the notion that a rocket could function in a vacuum and commented on the ideas of Robert H. Goddard, the rocket pioneer, as follows:

"That Professor Goddard, with his 'chair' in Clark College and the countenancing of the Smithsonian Institution, does not know the relation of action to reaction, and of the need to have something better than a vacuum against which to react—to say that would be absurd. Of course he only seems to lack the knowledge ladled out daily in high schools."

Further investigation and experiment have confirmed the findings of Isaac Newton in the 17th Century, and it is now definitely established that a rocket can function in a vacuum as well as in an atmosphere. The *Times* regrets the error.

The next day's issue of *Time* magazine also contained an article I had written especially for the occasion, and as this (needless to say) appeared in a much abbreviated form, I would like to give the full version here.

For thousands of years the moon has signified many things to mankind: a goddess, a beacon in the night sky, a celestial body, an inspiration to lovers, a danger to beleaguered cities, a symbol of inaccessibility—and finally, a goal.

In only ten years, this last image has become dominant, but the change has occurred with such explosive speed that most of the world has not yet made the necessary emotional and mental adjustments. The stunning impact of the first close-up photographs still seems only yesterday; last Christmas, the crew of Apollo 8 swept over the far side of the moon and sent greetings back to Earth, 240,000 miles distant. Now, even before the wonder of that event has abated, we are preparing to land.

There may be setbacks—perhaps even disasters—in the years ahead; it is unreasonable to suppose that the conquest of a new and strange environment will not demand its toll. But men have never hesitated to pay the price, in blood as well as treasure, of exploration and discovery. Nor will they hesitate now, as they stand, for the second time in a thousand years, on the frontiers of a new world.

Like all human achievements, travel to the moon will pass through three phases: impossible, difficult, easy. The parallel with the development of commercial aviation will be close, though the time scale may be longer because the challenge is so much greater. But it is naïve to imagine that lunar flight must always be an enormously expensive operation and that astronauts will always be highly trained pilots, scientists, or engineers.

If you run your car for a day, the engine does enough work to take you to the moon; the actual cost of the energy involved for the trip is only about ten dollars. The fact that the present cost is millions of times greater is the measure of our present ignorance and the primitive state of space technology; the time will come, through the use of reusable boosters, orbital refueling, nuclear propulsion, and other foreseeable developments, when the cost of a lunar journey may be comparable to that of round-the-world jet flight today.

It is obviously impossible, on the eve of the lunar landing, to predict in detail just what we shall do with an Africa-sized world, the resources of which are still almost entirely unknown. However, the moon provides such tremendous opportunities for so many types of research that every effort will be made to establish temporary bases there as soon as possible, analogous to those already set up in the Antarctic and those that may be established on the seabed.

Beyond the immediate deployment of small instrument packages that is planned on the Apollo missions, we may eventually expect physics laboratories and astronomical observatories. At first, they will be remote-controlled and visited from time to time by servicing crews; later, they will be permanently manned.

The moon might have been designed as the ideal site for an astronomical observatory. Its almost total absence of atmosphere means that seeing conditions are always perfect, not only in visible light, but also in the vitally important ultraviolet, X-ray, and gamma-ray regions of the spectrum, which are totally blocked by the earth's atmosphere. The low gravity and absence of wind forces will also greatly simplify the design of large instruments; and the slow rotation means that objects can be kept under continuous observation for two weeks at a time.

These advantages, great though they may be for the optical astronomer, will be even more overwhelming for the radio observer, who can also find another bonus on the moon. At the center of the far side, he will be permanently shielded from all the electrical noise and interference of civilization by two thousand miles of solid rock. A hundred years from now, optical and radio astronomers will find it hard to believe that serious observing was ever possible on Earth.

To the geologist, the moon represents a bonanza of more value than all the gold mines ever found. Until now, he has had a single example of a planet to study. How much would a biologist know of life, if he had been allowed to examine only one specimen of our planet's teeming flora and fauna?

The evolution and geological history of the moon may be wildly different from that of the earth; we are not even sure whether the two bodies were once combined or whether the moon had an independent origin and was later captured. One recent theory suggests that it is a residual "drop," a sort of umbilical fragment left over when the earth and Mars split asunder from an ancient protoplanet.

Whatever the facts, we can be sure that the moon will provide many exciting and valuable surprises. Indeed, it has already done so. In the astronomy books of only a decade ago, it was described as a dead, unchanged world. Now we know that there is a good deal of activity there. Orbiter photographs have shown the tracks of rolling rocks, startlingly like footprints, down (and sometimes up) the lunar slopes. There is evidence of immense lava flows, and even what looks like dried-up river valleys. If this is the case, water may still be there, locked in permafrost a few metres underground, where the temperature is constant and far below the freezing point.

The discovery of easily available water or ice would be of the greatest importance to lunar explorers. Electrolyzed, it would provide both oxygen for breathing purposes and fuel for returning spacecraft. Obviously, this last development would not be possible until large-scale engineering operations could be carried out on the moon. This is not likely for some decades, but eventually it will completely transform the economics of space flight. For a remote comparison, imagine that today's transatlantic aircraft had to carry the fuel they needed for the round trip. The cost of a ticket would be reduced by a factor of perhaps a hundred as soon as it became possible to refuel in Europe. So it will be with lunar operations.

After air and water, the third immediate necessity of life is food. Many plans have been drawn up for growing totally enclosed, or hydroponic, crops on the moon, using the materials that may be found there. This idea looks particularly promising, now that the Lunar and Surveyor spacecraft, in close-up views of the lunar surface, have revealed that it is neither rock nor dust, but nice, crumbly dirt.

Some years ago I suggested that it might be possible to develop plants resembling Earth's cacti with tough, impermeable skins that could grow unprotected on the lunar surface, and I am delighted to discover that the National Aeronautics and Space Administration now has a project investigating this idea. Perhaps a transparent plastic sheet may be necessary to minimize the escape of water vapor; but it is at least conceivable that we may start farming on the moon without having to build pressure domes and hermetically sealed greenhouses.

The lunar vacuum, so valuable to the astronomers, may turn out to be a much exaggerated hazard to the explorers. The old myth that a man exposed to the vacuum of space will blow up like a deep-sea fish still dies hard; I hope the movie *2001: A Space Odyssey* may have spread the news that this is simply not true. Obviously, an unprotected man in space will die from lack of oxygen, but this takes an appreciable time. Animals have survived up to four minutes in a vacuum, and anything an animal can do, a trained and prepared man can do better. There will be many emergencies, in space and on the moon, where the ten or fifteen seconds of consciousness that a man can expect in vacuum will make the difference between life and death.

Whether the moon has any indigenous life of its own is a question that may be answered shortly. No one expects to find higher organisms, but microscopic forms of life are a remote possibility. Hence the elaborate precautions of the Lunar Receiving Laboratory, which is intended to establish a quarantine in both directions.

Even if the moon is sterile, it may be avid for life. Those terrestrial bacteria that have managed to thrive in boiling sulphur springs or at the bottom of oil wells should find the moon a delightfully benign environment, with consequences that may be annoying to future scientists.

It has been estimated that the combustion products and cabin leakage from only twenty landings of the Apollo type could double the mass of the very tenuous lunar atmosphere. When mining, food production, and similar activities begin, the rate of contamination will be much increased and though it may seem early to worry about lunar smog, it could be a matter of great concern to the physicists.

At the moment, the moon's surface provides a vacuum laboratory of unlimited extent. It would be the ideal place for many types of electronic and nuclear experiments. One can even imagine that the great particle accelerators of the future will be wrapped around the moon, so that the vacuum will be provided automatically, and there will be no need for today's elaborate enclosures and pumps.

This sort of experimenting, which may well revolutionise the many branches of physics concerned with vacuum phenomena, may be possible only in the early stages of lunar occupation. For sooner or later, as industry, commerce, and tourism spread across the face of the moon, it will begin to acquire an atmosphere of its own.

And if it turns out, as some have suggested, that the expectation of life is considerably increased in low gravitational fields, there will be a move to give the entire moon a breathable atmosphere, probably by using biological systems to unlock the immense amounts of oxygen (about 50 percent by weight) bound up in the crust. The astronomers and physicists will have to move elsewhere in search of ideal conditions, just as on the surface of this planet they have had to retreat from the lights of the cities.

And a century or so after that, as I gloomily predicted in *The Promise of Space,* there will be committees of earnest citizens desperately trying to preserve the last vestiges of the lunar wilderness.

Not long ago, a critic of the space program suggested that as soon as the first astronauts came safely back from the moon, we should wind up the Apollo project and leave future exploration to robots.

This may well rank as the silliest statement of a notably silly decade; to match it one must imagine Columbus saying: "Well, boys, there's land on the horizon—now let's go home." Or better still—picture the Wright Brothers returning to the serious business of bicycle building as soon as they had demonstrated the trivial proposition that men could fly.

The moon is only the first milestone on the road to the stars. The exploration of space—by men and machines, for each complements the other—will be a continuing process with countless goals, but no final end. When our grandchildren look back at Earth, they will find it incredible that anyone there failed to realise so obvious a fact of life.

But it is not altogether obvious today, for belief in a space-faring future still requires certain elements of faith. One can understand the skepticism of those who point to the enormous cost (approximately $200 million) of an Apollo mission, and therefore see space travel as an expensive scientific stunt which even rich societies can only rarely afford.

This view of the future fails to recognize that today's space technology, for all its glittering hardware, is still in the log-canoe stage. Escaping from Earth will not always be astronomically expensive; contrary to the impression created by a Saturn launch, the energy needed to reach space is remarkably small.

About eight hundred pounds of kerosene and liquid oxygen, costing some twenty-five dollars, will liberate enough energy to carry a man to the moon. The fact that we currently burn a thousand tons per passenger indicates that there is vast room for improvement.

This will come, through space refuelling, nuclear propulsion and, most important of all, the development of reusable boosters, or "space ferries," which can be flown for hundreds of missions, like normal aircraft. We have to get away, as quickly as possible, from today's missile-orientated philosophy of rocket launchers which are discarded after a single flight.

When one considers how long it has taken to develop the supersonic transport, it can hardly be expected that the DC-3 of space will arrive before the 1980's. The next decade, therefore, despite all the spectacular achievements it will surely bring, will be a period of consolidation. Such a technological plateau occurred in 1945–1955, when the results of wartime rocket research had to be assimilated before the first breakthrough into space was possible. We are now entering a very similar period; sometime after 1985, the true Space Age will begin to dawn, and projects which today are barely feasible will become not only relatively easy, but *economically self-supporting*.

The closing years of this century should see the beginnings of commercial space flight, which will be directed first towards giant manned satellites or space platforms orbiting within a thousand miles of the earth's surface.

We can already envisage many industrial, scientific, and even medical uses for these; our astronauts have already shown us how the novel conditions of zero gravity can be enjoyed and exploited. Not only will new manufacturing and assembling processes be possible where all objects are weightless, but we may be able to create strange and useful materials (metal-air foams are one suggestion) impossible to make on Earth. It has even been suggested that one could "grow" giant diamonds in space—if anyone happens to want giant diamonds.

However, there is no need to think in terms of such exotic items as this. There are countless articles of commerce—drugs, the hairsprings of good watches, microelectronic components—whose cost per pound is so high that even today Earth-orbital freight charges would be unimportant. They will be quite trifling—perhaps less than ten dollars a pound—when large-scale manned operations commence.

And manned operation will be vital for the development of space industry. Even if, as is likely, most of the communications, meteorological, earth-resources survey, and other applications satellites will be automatic devices, as they grow larger and more complex we shall need human crews to install, service, and repair them. (There are about a billion dollars' worth of dead satellites in orbit at this moment, which might be repaired by a few men with screwdrivers.) For large space-factories and production plants, permanent onboard crews will be essential.

So even if the moon did not exist as a challenging target, all the necessary hardware for the manned exploration of the solar system would inevitably be developed by the scientific-industrial complex of the future. It would take much longer, of course, if we were foolish enough to wait until space operations were financially self-sustaining; cost-effectiveness is not a criterion that can, or should, be applied to advanced technology. (Who would have put money on atomic energy in 1940—even though the basic knowledge already existed then?)

Once we have the power to explore the moon and planets—and not merely to make brief reconnaissance raids—it is unthinkable that it will not be used. We can draw parallels here with the Antarctic and the oceans; at first they were barely accessible, but now we are establishing permanent bases there. Compared with both of these, the moon is a benign and stable environment, though because of its strangeness men have been slow to recognize this fact. It will not tolerate mistakes, but it does not possess the implacable hostility of the polar winter, or of the deep oceanic trenches. We can live, work, and even flourish there—if we wish to do so.

And will we? The answer to this question depends upon the resources we discover on the moon—not only in materials like oxygen, metal, and water but such imponderables as power, knowledge, and the capacity to do things that are not possible on Earth. Those who regarded the moon as a changeless cosmic slagheap have already been proved wrong; there are bigger surprises to come. It is inconceivable that a whole world, with a surface area larger than Africa, will not provide limitless opportunities for research, exploitation, and even colonisation. If it serves no other purpose, it may play a vital role as a refuelling stop in the opening up of the solar system; for it is twenty-five times easier to reach deep space from the surface of the low-gravity moon than from the earth. When the next century begins the conquest of the planets, our solitary natural satellite may be its launching pad.

In our present state of almost total ignorance, the only prediction that can be safely made about the other eight planets and their thirty-odd moons is that there is not a single one upon which unprotected men can live. Most of these places are almost unimaginably alien; but that very fact will give them immense scientific value. Moreover, simple arithmetic shows that if our present rate of industrial growth, power production, and use of raw material continues to increase, in a very short time—historically speaking—we will be forced to exploit the resources of other worlds. This does not give us a charter to continue turning Earth into a planetary garbage dump; we must put our own house in order before we expand into others. But it is nice to know that they are there— even though extensive alterations will be required to make them comfortable. Our generation has learned how to kill a world; the same powers can bring life to worlds that have never known it.

The real future is always much more astonishing than any prophet dares to predict. All our ideas about travel beyond the earth are based upon rocket technology, and indeed we have no other means of propulsion in the vacuum of space. But it may well be that the rocket's history will parallel that of the balloon, which lifted mankind into a new element, but was eventually superseded.

A Saturn V blast-off is the most magnificent spectacle contrived by man; yet there must be better ways of doing it, more compatible with nervous old ladies visiting their grandchildren on the moon, and with the peace and quiet of the countryside. The history of technology teaches us that the right tool always arrives at the right time; witness how the transistor was ready when the Space Age dawned.

It may be a pure coincidence, but there seems something uncanny in the fact that the long search for gravitational waves succeeded[1] only months before the first attempt to land on the moon. The old dream of controlling gravity may be a complete illusion—or it may foreshadow a basic industry of the twenty-first century. When Hertz proved the existence of electromagnetic waves, eighty years ago, he saw no practical use for them; now there is no man alive who is not affected by the communications networks which enmesh our globe.

The cycle may be beginning again, leading to feats of spacial engineering as inconceivable to us as television

---

1. This has now been questioned, but few scientists doubt that such waves *do* exist.

would have been to the Victorians. Whatever technologies the future may bring, the doors of heaven are now opening; this is the central fact of our age, and of all the ages to come.

Those who are—understandably—obsessed with the urgent problems of today aim at the wrong target when they attack the space program. A nation which concentrates on the present will have no future; in statesmanship, as in everyday life, wisdom lies in the right division of resources between today's demand and tomorrow's needs. It is true that when the house is on fire, the farmer must abandon the sowing; but he will lose more than his home if he does not continue to prepare for next season's harvest.

Before this year ends, decisions made by a handful of men will determine the future of many worlds. It will be ironic if the historians of the twenty-first century record that the solar system was lost in the paddy fields of Vietnam.

11

The Next Twenty Years

In 1972 the *Chicago Tribune Magazine* celebrated twenty years of its existence by publishing its one-thousandth issue, and I was requested to write an essay forecasting the events of the next twenty years. As I remarked in Chapter 8, I am usually reluctant to do this sort of thing; it's so much safer to write "The Next Million Years."

Anyway, here is the result, which makes interesting comparison with "The World of 2001," written for *Vogue* six years earlier. Avoiding duplication was such hard work that I have no intention of tackling this theme again until 2001—*if* then.

Until recently, everybody talked about the future, but no one did anything about it. Well, we are still talking—more so than ever—but there's also a lot of *doing*. This is the result of a quite remarkable change in our attitude towards the future.

It is hard to remember that, not long ago, "planning" was almost a dirty word to many people; some simple souls apparently thought that only communists had "Plans." But now we are seeing, on a horrendous scale, the effects of bad planning, and of no planning at all. Everyone can list examples of past short-sightedness—our decaying cities, poisonous rivers, unbreathable air, choked highways. . . . The whole ecology movement is a reaction (sometimes an overreaction) to the laissez faire of the past. It is now widely realised that we don't have to sit back and wait for the future to clobber us. To a considerable extent we can, in Nobel-laureate Dennis Gabor's striking phrase, *"invent* the future." In fact, it might well be argued that unless we invent a better future, we won't have one of any kind.

This is the first age in which man has had a major degree of control over his own destiny, for better or for worse. He has become a global force, already affecting the climate and even the habitability of his own planet. Of course, there are still many natural phenomena he can do nothing about, and which may always lie beyond his complete control. Hurricanes and earthquakes are the most spectacular examples, but even these may be one day fully predictable, and their worst consequences avoided.

Serious thinkers about the future are therefore no longer much interested in such statements as: "This is the way things *will* be in the year 1992." Rather, they are liable to say: "This is the way things *may be unless we do so-and-so.*" Much science-fiction writing (perhaps most of the best, such as *Fahrenheit 451, Brave New World*) has been of a cautionary nature. Orwell wrote *1984* in the desperate hope of awakening mankind from the nightmare he foresaw. He may have succeeded; to that extent he will have turned out to be a bad "prophet"—and no one would have been happier.

The most useful question we can ask ourselves about the year 1992 is, therefore: "Starting from where we are now, and making reasonable assumptions about future trends and discoveries, how can we make 1992 better than 1972?"

Twenty years is both a very long, and a very short, time. Most of the houses now in existence will (unfortunately) still be occupied in 1992. The road and rail systems will still be basically the same, though it is hoped by then a few American cities will have subways nearly as good as those which Parisians and Londoners have enjoyed for a couple of generations. The 747's and 707's—or slightly stretched versions—will still be carrying a substantial fraction of the world's travellers. (And who would care to bet that there won't be a few DC-3's operating somewhere?) These features of our civilisation involve such gigantic investments of capital, material, and labor that, even with the strongest incentives, it requires decades to make substantial changes in them.

The jet revolution is a perfect example. It took the airlines about fifteen years to convert from props to jets, and the process is still not 100 percent complete. If someone came out tomorrow with a cheap, crash-proof car that ran on water—well, there would still be *some* gas stations in business even in 1992. There would not have been sufficient time to retire all the current models, or earn the money to pay for their replacements.

Yet in other fields, where large amounts of capital and material are not involved, it is possible to have several complete revolutions in twenty years. This is particularly true in life styles, art, politics, and social attitudes, where there can sometimes be a total transformation literally overnight. There has been ample proof of that in the last few years; who would have dared to predict, as recently as 1965, that by the beginning of the seventies the President of the United States would visit Peking, the American Medical Association would be debating the legality of marijuana, abortion would be respectable, total stage and screen nudity would have started to become boring . . . and the yen would be mightier than the dollar?

Some of these changes were totally unpredictable because, like women's fashions, they depend on factors that are largely arbitrary. Yet others might have been anticipated; it should have been obvious, for example, that increasing population pressure would produce changes in sexual morality.

If I may be allowed the modest cough of the minor prophet, I would like to recall a couple of predictions I made in this area, just about twenty years ago. The novel *Childhood's End* (1953) described a society in which mores had been totally transformed by two inventions. The first was an oral contraceptive; the second was a technique "as infallible as fingerprinting" for identifying the father of any child. Well, the first invention has already arrived; by 1992 we may have the second. But I am no longer sure if it is important; the liberated woman of that day may not give a damn.

Twenty years from now humanity will be in the midst of one of its most painful and difficult social changes. This century will be the last in which families of more than two children can be tolerated; everyone knows this, and the only argument is over the means of achieving the goal. But there is another aspect of the matter which is seldom given much serious consideration.

The two-child family is not large enough to generate the interactions that develop a good personality; this is why single children are often monsters. Probably the optimum number of siblings is four or five—twice the permissible quota. This means that, somehow, several families must be psychologically fused together for the health of the child, and of society. Working out ways of doing this will raise the blood pressure of a whole generation of lawyers and moralists.

It is also probable that by 1992 the present furore over drugs will seem as remote as the very similar hysteria over Prohibition does today. An increasingly mature society will recognise (as indeed it is already doing) that legislation of morality is not only impossible, but counterproductive—and therefore in the long run essentially *immoral*. People should be encouraged not to commit suicide with alcohol, tobacco, or heroin, but if they are sufficiently determined they should perhaps be helped to get on with the job. However, it is not obvious where one draws the dividing line between free whiskey, free cigarettes, and free cocaine. . . .

No wonder that so many discussions of the future avoid

the sticky and messy issues of human behavior and the structure of society, and concentrate on technology. It's a lot easier—and frankly more exciting—to talk about new inventions and scientific achievements than the problems of civilising the cities, or salvaging the waste products of our so-called educational system. Yet this "Buck Rogers Syndrome," as it might be called, is not to be despised; a new technology *can* transform society, as Henry Ford convincingly demonstrated.

It is therefore extremely important to identify the real breakthroughs in invention—not the Madison Avenue ones, of which the mythical (I hope) power zipper will serve as prototype. The really crucial invention is one that affects every aspect of human life, far outside its own obvious field. Television is a particularly good example; at first it was conceived merely as a form of luxury entertainment, but within twenty years it had become a dominant factor in sports, news, and—above all—politics. It is just beginning to make its impact on education.

The story of TV contains valuable warnings and lessons. In the 1940's, one of the most far-sighted men of his time, the late John W. Campbell, science-fiction writer and editor, argued convincingly that TV would never come into wide domestic use. A piece of equipment as complicated, fragile, and expensive as a TV set simply could not be made and sold in the millions.

Campbell failed to take into account the insatiable demands of human beings for information and entertainment. The TV set could provide these in abundance, beyond the wildest dreams of the past. To every home, it gave a magic window on the world. Men can never resist magic, and so shanties without plumbing sprouted antennas.

In just twenty years, TV had covered all the so-called developed countries; the last holdout, South Africa, tried to resist for ideological reasons, but in the end even that conservative society was forced to capitulate, despite the extremely perceptive warnings of Afrikaners who pointed out that TV would inevitably mean the end of white dominance. By 1992 there will be no spot on Earth where good TV cannot be received by cheap, simple equipment, and the human race will be plugged into a single communications system.

This electronic revolution is the one through which we are now passing; it has already transformed our lives (can you remember the world before TV?) but the greatest changes are still to come. Though they will not be complete by the 1990's, by that time the shape of a new future will be starting to emerge—if we desire it sufficiently.

It will be a future in which men do much less commuting, much more communicating. Even today, probably 90 percent of the average executive's business could be performed *without leaving* home, by the use of equipment which is already available on an experimental basis. During the next decade, we will see the evolution of a general-purpose, home-communications console providing two-way vision, hard-copy readout so that diagrams and printed material can be exchanged, and a keyboard to allow "conversation" with the computers and information banks upon which our world will increasingly depend.

Before we consider its practicability, let us see what we could do with such a device. Far more than business discussions and conferences would be possible; the housewife could go shopping by dialing the catalogues of her favorite stores; scholars and students would have instant access to any book or periodical stored in the global electronic library; this minute's news, continually updated, would be displayed in printed headlines, and any selected item could be expanded as desired, according to taste. This, incidentally, raises the possibility of something quite new—the "personalized" electronic news service, tailored to the interests of the individual subscriber!

Today, such a receiving console would cost tens of thousands of dollars—and would be useless, because the communications network to service it does not yet exist.

But this network will be built up during the next decades; one of the great enterprises of the twentieth century will be the establishment, with the help of satellites, of a planetary "information grid." It will join the other networks we have developed during the last hundred and fifty years, and which we now take so much for granted that we forget their existence, except when they break down. Chronologically, they are: water, sewerage, gas, electricity, telephone—and now cable TV, or video. The forthcoming information grid will absorb the last two.

How long will this take, and is it economically and technically feasible? The speed of the TV revolution suggests that twenty years from the first demonstration is a reasonable guess; this would put the general use of home consoles in the 1990's. As for the cost factor, even if the console was as expensive as a small car (which it might be in the early days) this would not inhibit its wide-scale use. After all, who would have dreamed, at the beginning of this century, that there would be a car for every home? Eventually, the console would cost no more than the TV set of 1972. Anyone who doubts this should look at the history of computers during the last quarter of a century.

The first genuine electronic computer, the 1946 ENIAC, contained eighteen thousand vacuum tubes (remember them?), was the size of a small house, used kilowatts of power, and probably cost a million dollars. Today's "electronic slide rules" contain far more than eighteen thousand of the solid-state elements that have replaced vacuum tubes—but fit neatly in the palm of the hand, run for hours on their internal batteries, and cost a hundred dollars.

And here is another statistic, in some ways even more impressive. I quote from Dennis Gabor's stimulating book, *Innovations:*

> Around 1950, some very able people had estimated that the whole computation work in the United States could be carried out by a dozen of the (still

rather slow) electronic computers which existed at that time. The estimate for Britain was *two* computers. Never before have such able people underestimated a market by a factor of at least 10,000.

The communications revolution may also proceed at a rate that will make the experts look foolish. Unlike the rebuilding of cities or the construction of superhighways, it requires relatively modest amounts of raw material and exceedingly small amounts of power. The electronic information industry does not pollute the environment (whatever it may occasionally do to the mind!). As Buckminister Fuller has pointed out, communications satellites are an ideal demonstration of his principle of "doing more with less." A one-ton satellite can replace hundreds of thousands of tons of cables and microwave towers; in a more indirect way, it may also replace thousands of miles of roads.

What we may see coming, therefore, is a world in which routine, daily commuting is vastly reduced, but travel for pleasure is increased. Even when every business and professional activity can be carried out by electronic means, people will still wish to get together; men are social animals—witness the crowds at sporting events which can be much better observed on TV. Thus we may expect the phasing-out of the commerce-directed megalopolis and the rise of much smaller complexes for entertainment, culture, sports. The prototype of the future city is not Manhattan, but Disney World. . . .

This transformation will scarcely have started by 1992, but its possibility may be so obvious by then that it will unfortunately discourage those who are attempting to save the existing cities, which will be so clearly obsolescent.

Something else which may be on its way out at that date is agriculture as we know it—particularly meat production. The so-called Green Revolution—the greatest in the entire history of agriculture—occurred in the last twenty years; a still greater one is imminent.

The manufacture of meat from sunlight, air, water,

and soil, which is what conventional farming is all about, is an unbelievably inefficient process, involving vast acreages of land. It can be short-circuited if we go straight to oil as raw material; at this moment, in France, Japan, and elsewhere, the first factories are already manufacturing protein from oil. By the mid-seventies the Japanese hope to be producing a million tons a year. When this process is perfected, it is believed that synthetic meat— indistinguishable from the natural product in taste, appearance, *and* nutritive value—will be available for a fraction of the price the housewife has to pay today. This will be a revolution indeed, which should also please all those who object to killing animals for food.

But there is a problem here. If we stopped burning gas and started eating it, there would be enough to feed mankind for centuries—perhaps millennia. However, eventually the basic material would be exhausted.

Agricultural techniques, on the other hand, are self-regenerating, if properly carried out, which is often far from the case. Good farmers will never starve in a million years, because they put back into the earth all that they get out of it. But we are going to use up this planet's oil reserves anyway; better that they fill our bellies than poison our lungs. . . .

The year 1972 was the one in which, culminating with the United Nations conference in Stockholm, the attention of the world was focussed on the twin problems of resources and pollution. (Actually, all pollution is simply an unused resource.) Around 1992 we shall know if the grimmer prophecies of the pessimists have been realised. If they have not, it will be because our disposable, throw-away economy has come to its senses. By that date, it should be recognised that it is a social crime to tie up valuable materials, labor, and energy in a product that is designed for the rubbish dump in five years. There will be many items coming into use which will be guaranteed for ten, twenty, fifty years—even for the lifetime of the purchaser. This will alleviate the present wastage of raw materials, but we will still need vast new supplies of metals, minerals, and power if the standard of living of all mankind is to be brought up to a civilised level. Although the resources of this planet are certainly limited, reports of their imminent exhaustion are greatly exaggerated. We have just discovered that there are metal deposits on and under the seabed that may dwarf those on land; the technical problems involved in reaching them may be easier to solve than the legal ones of deciding their ownership. Just as the oil industry has moved out to sea in the last twenty years, not always with happy results, so we may expect in the next twenty the establishment of massive underwater mining operations. One very promising area—if politics allows—could be the Red Sea. Its depths are literally oozing with hot brines loaded with heavy metals, perhaps laying down the ore beds of the year 200,000,000. We cannot wait until then.

The other much-feared shortage of the future is that of power, and in the next twenty years we will have to tap alternative sources of energy. One which has been largely neglected, except in a few volcanic regions, is geothermal power—the almost inexhaustible outpouring of heat from the interior of the earth. In principle, it would appear that we have only to sink a couple of deep shafts, pump cold water down one and let steam come up the other—and we would have clean, cheap power for an indefinite period. Many people believe that this is exactly what we will be doing within the next few years.

The last twenty years have seen the economical harnessing of atomic energy—as great a revolution as any that man has ever experienced, and one fraught with frightful peril as well as promise. Today's nuclear power stations all depend upon the splitting of the heavy metal uranium, and their unavoidable waste products include embarrassing amounts of deadly radioactive materials.

This is one reason why a great deal of effort (though not enough) has been expended in the search for "fusion power," the process by which the sun squeezes together hydrogen atoms to make heavier elements. So far, we have succeeded only in starting this reaction explosively, in the hydrogen bomb; to do it on a slow and controllable scale is much more difficult. But when it is achieved, we shall have a virtually limitless source of power and, equally important, it will be almost clean.

The breakthrough—and here the word is for once fully justified—may come tomorrow, or it may come ten years from now. I shall be very surprised (and disappointed) if it has not come by 1992. I shall also be very surprised (and delighted) if there are any fusion power stations by that date. But they will probably be on the drawing board, and we will know that our energy problems are over, for as far ahead as imagination can see.

It is already very hard to believe that the last twenty years more than spans the whole of the Space Age, ushered in with Sputnik in 1957. When that hundred-pound sphere went beeping into orbit, not even the wildest optimist would have predicted that within a mere twelve years men would be walking on the moon.

Twenty years from now, men will be living there, and the first child may well have been born in some lunar colony. But more important from the point of view of earth-dwellers will be activities in near orbit; by 1992 much of the day-to-day running of our world will utterly depend on various types of specialised satellites. Their impact upon communications and weather forecasting is already profound, and 1972 saw the launching of the first Earth Resources Satellite, which will ultimately allow us to inventory our planet's natural wealth with a speed and economy impossible by any other means. In another generation it will seem incredible that we were ever able to run our world efficiently without the new tools of space, and today's critics of the space effort will seem as short-sighted as those who laughed at automobiles.

And by 1992, if the first men are not on Mars, they will be preparing to go there. It was in 1972 that we discovered our mysterious and romantic little neighbor was almost as exciting as some of the science-fiction writers had imagined. But the exploration—and colonisation—of Mars belongs on the far side of 2001; it may well be one of the dominant themes of the twenty-first century.

One safe prediction about the twenty years that lie ahead is that they will see at least one development or discovery which no one could have predicted. (The laser is a good example from the recent past.) Will the year 1992 see something as wholly unexpected as, for example, atomic energy or relativity was to the physicists at the beginning of this century? What could it possibly be. . . ?

And therein lies much of the future's fascination; we know that it will always contain some element of the utterly unforeseen. Indeed, if we could *really* anticipate the future in every detail, that would be a disaster. We would lose even the illusion of free will; nothing that we could do would ward off misfortune, if we predicted it correctly! And even good fortune would no longer be enjoyed, for all pleasures would be dissipated in their anticipation.

It is sometimes charged that thinking about the future is a form of escapism—an attempt to avoid the problems of the present. We should not take this accusation too seriously, even when there is some truth in it; who has not, at one time or other, indulged in that wistful dreaming that has been called "nostalgia for the future"?

Such nostalgia is harmful only when it leads to fatalism and inhibits action. The two decades from 1972 to 1992 present nightmare problems, but also tremendous opportunities. How we face them will determine not only whether we will survive but whether we *deserve* to survive.

12

Satellites and Saris

On 20 August 1971, after several years of Byzantine negotiations, the final agreements setting up the world satellite communications system (Intelsat) were signed at the State Department in Washington. At the invitation of Secretary William Rogers and Ambassador Abbott Washburn, the United States representative to Intelsat, I was asked to speak at the ceremony, immediately following William Anders, then executive secretary of the National Aeronautics and Space Council.

I have never quite forgiven Bill Anders for his failure of nerve on Christmas Day, 1968. He had been tempted, he later told me, to radio back from Apollo 8 that the crew had spotted a large black monolith on the far side of the moon.

In my Intelsat address, I referred to the forthcoming Indian educational satellite experiment, then still almost four years in the future. The article that follows, "Schoolmaster Satellite," was an attempt to explain the purpose —and the high hopes—of this great experiment, in many ways the most promising of all the applications of space technology. It was later read into the *Congressional Record* (27 January 1972) by Representative William Anderson, who skippered the atomic submarine *Nautilus* to the North Pole in 1958. When I wrote the article, which also appeared in the London *Daily Telegraph* colour magazine for 17 December 1971, I could not have guessed how closely this experiment would affect my own life. But of that, more in Chapter 23.

Mr. Secretary, your excellencies, distinguished guests. . . . Whenever I peer into my cloudy crystal ball and try to visualise the future of communications satellites, I remember an incident that occurred in England almost a hundred years ago.

The very alarming news had just been received from the United States that a certain Mr. Bell had invented the telephone. This, of course, was very disturbing. So, as we British do in an emergency, we called a parliamentary commission. It listened to the evidence of expert witnesses, who gave the reassuring news that nothing further would be heard of this impractical Yankee invention.

Among the witnesses called was the chief engineer of the British Post Office. Someone on the commission said to him: "We understand that the Americans have invented a machine that can transmit human speech. Do you think that this—*telephone*—will be of any use in Great Britain?" The chief engineer thereupon replied: "No, sir. The Americans have need of the telephone, but we do not. *We* have plenty of messenger boys."

This very able man totally failed to see the possibilities of the telephone, and who can blame him? Could *anyone*, back in 1880, have imagined that the time would come when every home would have a telephone, and business and social life would depend upon it almost completely?

I submit, gentlemen, that the eventual impact of the communications satellite upon the whole human race will be at least as great as that of the telephone upon the so-called developed societies. In fact, as far as real communications are concerned, there are as yet no developed societies; we are all in the semaphore and smoke-signal stage. And we are now about to witness an interesting situation in which many countries—particularly in Asia and Africa—are going to leapfrog a whole era of communications technology and go straight into the Space

Age. They will never know the vast networks of cables and microwave links that this continent has built up at such enormous cost. Satellites can do far more, at far less expense.

Intelsat, of course, is concerned primarily with point-to-point communications involving large ground stations, often only one per country. It provides the first reliable, high-quality, wide band with links between all nations that wish to join, and the importance of this cannot be overestimated. Yet it is only a beginning, and I would like to look a little further into the future. . . .

Two years from now, NASA will launch the first satellite—ATS-6—which will have sufficient power for its signals to be picked up by an ordinary domestic TV set, plus about two hundred dollars' worth of additional equipment. In 1974[1] this satellite will be stationed over India and, if all goes well, the first experiment in the use of space communications for mass education will begin.

I have just come from India, where I have been making a TV film, *The Promise of Space*. We erected, in a village outside Delhi, the prototype antenna: a simple umbrella-shaped wire-mesh affair, three metres across. Anyone can put it together in a few hours; it needs only one per village to start a social and economical revolution.

The *engineering* problems of bringing education, literacy, improved hygiene, and agricultural techniques to every human being on this planet have now been solved. The cost would be of the order of a dollar per person— *per year*. The benefits in health, happiness, and wealth would be immeasurable.

But, of course, the technical problem is the easy one. Do we have the imagination—and the statesmanship— to use this new tool for the benefit of all mankind? Or will it be used merely to peddle detergents and propaganda?

I am an optimist; anyone interested in the future has to be, otherwise he would simply shoot himself. . . . I

---

1. The program actually began in August 1975.

believe that communications satellites can unite mankind. Let me remind you that this great country was virtually created a hundred years ago by two inventions. Without them, the United States was impossible; with them, it was inevitable. Those inventions were, of course, the railroad and the electric telegraph.

Today we are seeing, on a global scale, an almost exact parallel to that situation. What the railroads and the telegraph did here a century ago, the jets and the communication satellites are doing now to all the world.

I hope you will remember this analogy in the years ahead. For today, gentlemen, whether you intend it or not, whether you *wish* it or not—you have signed far more than yet another intergovernmental agreement.

You have just signed the first draft of the Articles of Federation of the United States of Earth.

For thousands of years, men have sought their future in the starry sky. Now this old superstition has at last come true, for our destinies do indeed depend on celestial bodies—those that we have created ourselves.

Since the mid-sixties, the highly unadvertised reconnaissance satellites have been quietly preserving the peace of the world, the weather satellites have guarded millions against the furies of Nature, and the communications satellites have acted as message-carriers for half the human race. Yet these are merely the first modest applications of space technology to human affairs; its real impact is still to come. And, ironically, the first country to receive the benefits of space *directly* at the home and village level will be India, where, as recently as February 1962, millions were terrified by an unusual conjunction of the sun, moon, and five planets.

In 1975 there will be a new Star of India; though it

will not be visible to the naked eye, its influence will be greater than that of any zodiacal signs. It will be the satellite ATS-6 (Applications Technology Satellite 6), the latest in a very successful series launched by the National Aeronautics and Space Administration. For one year, under an agreement signed on 18 September 1969, ATS-6 will be loaned to the Indian government by the United States, and will be "parked" thirty-six thousand kilometres above the equator. At this altitude it will make one revolution every twenty-four hours and will therefore remain poised over the same spot on the turning earth; in effect, India will have a TV tower thirty-six thousand kilometres high, from which programs can be received with almost equal strength over the entire country.

Since the launch of the historic Telstar in 1962, there have been several generations of communications satellites. The latest, Intelsat IV, can carry a dozen TV programs or up to nine thousand telephone conversations across the oceans of the world. But all these satellites have one thing in common; their signals are so feeble that they can be received only by large earth stations, equipped with antennas twenty or more metres across, and costing several million dollars. Most countries can afford only one such station, and indeed that is all that they need to connect their television, telephone, or other services—where these exist—to the outside world.

ATS-6, built by the Fairchild Corporation, represents the next step in the evolution of communications satellites. Its signals will be powerful enough to be picked up, not merely by multimillion-dollar earth stations, but by simple receivers, costing two or three hundred dollars, which all but the poorest communities can afford. This level of cost would open up the entire developing world to every type of electronic communication, not only TV. The emerging societies of Africa, Asia, and South America could thus bypass much of today's ground-based technology, and leap straight into the Space Age. Many of them have already done something similar in the field of transportation, going from oxcart to airplane with only a passing nod at cars and trains.

It can be difficult for those from nations that have taken a century and a half to slog from semaphore to satellite to appreciate that a few hundred kilograms in orbit can now replace the continentwide networks of microwave towers, coaxial cables, and ground transmitters that have been constructed during the last generation. And it is perhaps even more difficult, to those who think of TV exclusively in terms of old Hollywood movies, giveaway contests, and soap commercials, to see any sense in spreading these boons to places which do not yet enjoy them. Almost *any* other use of the money, it might be argued, would be more beneficial.

Such a reaction is typical of those who come from developed (or overdeveloped) countries, and who accept libraries, telephones, cinemas, radio, TV, as part of their daily lives. Because they frequently suffer from the modern scourge of information pollution, they cannot imagine its deadly opposite: information starvation. For any Westerner, however well-meaning, to tell an Indian villager that he would be better off without access to the world's news, knowledge, and entertainment is an impertinence. A fat man preaching the virtues of abstemiousness to the hungry would deserve an equally sympathetic hearing.

Those who actually live in the East, and know its problems, are in the best position to appreciate what cheap and high-quality communications could do to improve standards of living and reduce social inequalities. Illiteracy, ignorance, and superstition are not merely the results of poverty, they are part of its cause, forming a self-perpetuating system which has lasted for centuries, and which cannot be changed without fundamental advances in education. India is now beginning a Satellite Instructional Television Experiment (SITE) as a bold attempt to harness the technology of space for this task; if it succeeds, the implications for all developing nations will be enormous.

SITE's first order of business will be instruction in family planning, upon which the future of India (and all other countries) now depends. Puppet shows are already being produced to put across the basic concepts; those of us who remember the traditional activities of Punch and Judy may find this idea faintly hilarious. However, there is probably no better way of reaching audiences who are unable to read, but who are familiar with the travelling puppeteers who for generations have brought the sagas of Rama and Sita and Hanuman into the villages.

Some officials have stated, perhaps optimistically, that the only way in which India can check its population explosion is by mass propaganda from satellite—which alone can project the unique authority and impact of the TV set into every village in the land. If this is true, we have a situation which should indeed give pause to those who have criticised the billions spent on space.

The emerging countries of what is called the Third World may need rockets and satellites much more desperately than the advanced nations which built them. Swords into ploughshares is an obsolete metaphor; we can now turn missiles into blackboards.

Next to family planning, India's greatest need is increased agricultural productivity. This involves spreading information about animal husbandry, new seeds, fertilisers, pesticides, and so forth; the ubiquitous transistor radio has already played an important role here. In certain parts of the country, the famous "miracle rice" strains, which have unexpectedly given the whole of Asia a few priceless years in which to avert famine, are known as "radio paddy," because of the medium through which farmers were introduced to the new crops. But although radio can do a great deal, it cannot match the effectiveness of television; and of course there are many types of information that can be fully conveyed only by images. Merely *telling* a farmer how to improve his herds or harvest is seldom effective. But seeing is believing, if he can compare the pictures on the screen with the scrawny cattle and the dispirited crops around him.

Although the SITE project sounds very well on paper, only experience will show if it works. The "hardware" is straightforward and even conventional in terms of today's satellite technology; it is the "software"—the actual programme—that will determine the success or failure of the experiment. In 1967 a pilot project was started in eighty villages round New Delhi, which were equipped with television receivers tuned to the local station. (In striking contrast to a satellite transmitter, this has a range of only twenty-five miles.) It was found that an average of four hundred villagers gathered at each of the evening "tele-clubs," to watch programmes on weed control, fertilisers, packaging, high-yield seeds—plus five minutes of song and dance to sweeten the educational pill.

We who are accustomed to individual or family viewing tend to forget that even a twelve-inch set can be seen by several hundred people. Moreover, as it is always dark in India by 7 P.M., for much of the year the receiver can be set up in the open air; only during the monsoon would it be necessary to retreat into a village hall.

Surveys have been carried out to assess the effectiveness of these programmes. In the area of agricultural knowledge, TV viewers have shown substantial gains over nonviewers. To quote from the report of Dr. Prasad Vepa of the Indian National Committee for Space Research: "They expressed their opinion that the information given through these programmes was more comprehensive and clearer compared to that of the other mass media. Yet another reason cited for the utility of TV was its appeal to the illiterate and small farmers *to whom information somehow just does not trickle*" (my italics).

In February 1971, while filming *The Promise of Space*, I visited one of these TV-equipped villages: Sultanpur, a prosperous and progressive community just outside Delhi, only a few miles from the soaring sandstone tower of the Kutb Minar. Dr. Vikram Sarabhai, chairman of the Atomic Energy Commission, had kindly lent us a prototype of the three-metre-wide, chicken-wire receiving dish which will collect signals from ATS-6 as it hovers above the equator. While the village children watched, the pie-shaped pieces of the reflector were assembled: a job that can be performed by unskilled labour in a couple of hours. When it was finished, we had something that looked like a large aluminium sunshade or umbrella with a collecting antenna in place of the handle. As the whole assembly was tilted up at the sky and lifted on to the roof of the highest building, it looked as if a small flying saucer had swooped down upon Sultanpur.

With the Delhi transmitter standing in for the still un-launched satellite, we were able to show a preview of—one hopes—almost any Indian village of the 1980's. The programme we actually had on the screen at Sultanpur was a lecture-demonstration in elementary mechanics, which could not have been of overwhelming interest to most of the audience; nevertheless, it seemed to absorb viewers whose ages ranged from under ten to over seventy. Yet it was not at Sultanpur, but over six hundred kilometres away at Ahmedabad, that I really began to appreciate what could be done through even the most elementary education at the village level.

Near Ahmedabad is the big parabolic dish, sixteen metres in diameter, of the experimental satellite communications ground station, through which the programmes will be beamed up to the hovering satellite. Also in this area is AMUL, the largest dairy cooperative in the world, to which more than a *quarter of a million* farmers belong. After we had finished filming at the big dish, our camera team drove out to the AMUL headquarters, and we accompanied the chief veterinary officer on his rounds.

At our first stop, we ran into a moving little drama that we could never have contrived deliberately, and which summed up half the problems of India in a single episode. A buffalo calf was dying, watched over by a tearful old lady who now saw most of her worldly wealth about to disappear. If she had called the vet a few days before—there was a telephone in the village for this very purpose—he could easily have saved the calf. But she had tried charms and magic first; they are not *always* ineffective, but antibiotics are rather more reliable. . . .

I will not quickly forget the haggard, tear-streaked face of that old lady in Gujarat; yet her example could be multiplied a million times. The loss of real wealth throughout India because of ignorance and superstition must be staggering. If it saved only a few calves per year, or increased productivity only a few percentage points, the TV set in the village square would quickly pay for itself. The very capable men who run AMUL realise this; they are so impressed by the possibilities of TV education that they plan to build their own station to broadcast to their quarter of a million farmers. They have the money, and they cannot wait for the satellite, though it will reach an audience two thousand times larger, for over *500 million* people will lie within range of ATS-6.

There is a less obvious, yet perhaps even more important, way in which the prosperity and sometimes very existence of the Indian villagers will one day depend upon space technology. The life of the subcontinent is dominated by the monsoon, which brings 80 percent of the annual rainfall between June and September. The date of onset of the monsoon, however, can vary by several weeks, with disastrous results to the farmer if he mistimes the planting of his crops.

Now, for the first time, the all-seeing eye of the meteorological satellites, feeding information to giant computers, gives real hope of dramatic improvements in weather forecasting. But forecasts will be no use unless they get to the farmers in their half a million scattered villages and, to quote from a recent Indian report:

This cannot be achieved by conventional methods
of telegrams and wireless broadcasts. Only a space
communications system employing TV will be . . .
able to provide the farmer with something like a
personal briefing. . . . Such a nation-wide rural TV
broadcasting system can be expected to effect an increased agricultural production of at least 10 per
cent through the prevention of losses—a savings of
$1,600 million per annum.

Even if this figure is wildly optimistic, it appears that the
costs of such a system would be negligible compared to
its benefits.

And those who are unimpressed by mere dollars should
also consider the human aspect—as demonstrated by the
great Bangladesh cyclone of 1971. *That* was tracked by
the weather satellites, but the warning network that
might have saved several hundred thousand lives did not
exist. Such tragedies will be impossible in a world of
efficient space communications.

Yet it is the quality, not the quantity, of life that
really matters. Men need information, news, mental
stimulus, entertainment. For the first time in five thousand years, a technology now exists that can halt and
perhaps even reverse the flow from the country to the city.
The social implications of this are profound; already, the
Canadian government has discovered that it has to
launch a satellite so that it can develop the Arctic. Men
accustomed to the amenities of civilisation simply will not
live in places where they cannot phone their families or
watch their favourite TV show. The communications
satellite can put an end to cultural deprivation caused by
geography. It is strange to think that, in the long run, the
cure for Calcutta (not to mention London, New York,
Tokyo) may lie thirty-six thousand kilometres out in
space.

The SITE project will run for one year, and will broadcast to about five thousand TV sets in carefully selected

areas. This figure may not seem impressive when one considers the size of India, but it requires only one receiver to a village to start a social, economic, and educational revolution. If the experiment is as great a success as Dr. Sarabhai and his colleagues hope (and deserve), then the next step would be for India to have a full-time communications satellite of her own. This is, in any case, essential for the country's internal radio, telegraph, telephone, and Telex services.

It may well be that, until it has established such a nationwide system, India will be unable to achieve a real cultural identity, but will remain merely a collection of states. And one may wonder how much bloodshed and misery might have been avoided had the two severed wings of Pakistan been able to talk to each other face to face, through the facilities which only a communications satellite can provide.

Kipling, who wrote a story about "wireless" and a poem to the deep-sea cables, would have been delighted by the electronic dawn that is about to break upon the subcontinent. Gandhi, on the other hand, would probably have been less enthusiastic; for much of the India that he knew will not survive the changes that are now coming.

One of the most magical moments of Satyajit Ray's exquisite *Pather Panchali* is when the little boy Apu hears for the first time the Aeolean music of the telegraph wires on the windy plain. Soon those singing wires will have gone forever; but a new generation of Apus will be watching, wide-eyed, when the science of a later age draws down pictures from the sky—and opens up for all the children of India a window on the world.

13

The Sea of Sinbad

Most of my literary and artistic friends seem to have been first encountered in the lobby or elevator of the Hotel Chelsea, which is one of the reasons why it has been my New York headquarters for more than twenty years. I just missed Dylan Thomas and Brendan Behan, but the characters I have met include Arthur Miller; Paul Bowles; Andy Warhol (and the girl who shot him); Alan Watts; William Burroughs; Allen Ginsberg; Gregory Corso; Norman Mailer; Virgil Thomson, the composer; George Kleinsinger (with his famous "zoo" on the tenth floor); film producers Shirley Clarke, Conrad Rooks, and Milos Forman; sculptor René Shap-Shak; Charles Jackson (who was never able to repeat the success of *Lost Weekend*); *Hair*'s creators Jerry Ragni and Jimmy Rado; and, perhaps most memorable of all, Edith and Cliff Irving, at the height of their Howard Hughes escapade. If from this you deduce that the clientele is, perhaps, a little on the bohemian side, you are quite entitled to your opinion. It is not true, however, that the only house rule enforced by manager Stanley Bard is "Bring out your own dead." He takes a poor view of private bomb factories or window-box marijuana plots, and has been known to speak quite severely to residents who were only a few months in arrears with their rent.

Mere film producers and photographers are two a penny in this hotbed of genius (and occasional talent). But I was much impressed by the work of a young American, Peter Castellano, who eventually bought his Nikons and Hasselblads to Sri Lanka and did picture stories about the island for British Airways and the London *Observer* colour magazine. My prose can hardly match the glorious hues of Peter's photographs, but the two essays I wrote to go with them will, I hope, help to answer some of the questions I am always being asked. Including, as you will see from the second piece, some of the questions I had asked myself.

The island of Ceylon is a small universe; it contains as many variations of culture, scenery, and climate as some countries a dozen times its size. What you get from it depends on what you bring; if you never stray from your hotel bar or the dusty streets of Westernised Colombo, you could perish of fulminating boredom in a week, and it would serve you right. But if you are interested in people, history, nature, and art—all the things that *really* matter—you may find, as I have, that a lifetime is not enough.

Of course, not everybody feels that way; the writer Eric Linklater once left in disgust, rashly announcing to the world that "Ceylon stinks." The Ceylonese have carefully mispronounced his name ever since. Other distinguished visitors have been more flattering; Marco Polo declared that Ceylon is "undoubtedly the finest island of its size in all the world," and one could dig up similar quotations throughout the ages. The most beautiful tribute ever paid, however, must be that by a papal legate six centuries ago: "From Ceylon to Paradise, according to native tradition, is forty miles; there may be heard the sound of the fountains of Paradise."

You certainly won't hear the fountains if you arrive when the monsoon breaks, and the rain comes down in the best Somerset Maugham tradition. The finest time of the year for the western (Colombo) side of the island is December through March; there may be occasional showers, but most days are sunny and temperate, with the thermometer around 80° F. (27° C.). Between April and July it becomes humid and generally unpleasant in the west, but on the other side of Ceylon, behind a barrier of mountains up to two thousand metres high, the weather will be fine. Thanks to these mountains, there are two climates in a country less than three hundred kilometres across. By careful choice of time *and* place (and travel agent!) it is possible to enjoy Ceylon in any month of the year.

One of the factors that adds immeasurably to the convenience of visiting Westerners is that—in contrast to many exotic countries—English is spoken everywhere, and practically all street signs are in Roman characters (as well as in Sinhalese and Tamil). All educated Ceylonese are bilingual, since the country was a British colony until 1948. It must be admitted that, as a result of a drive to make Sinhalese the national language, the standard of English among the younger generation has plummeted during the last few years. However, attempts are now being made to correct this, and it would take a fairly determined traveller to find a place where not a single person understood that famous Victorian phrase-book appeal: "Help! My postillion has been struck by lightning."

The British influence, though waning, remains powerful; one of its most obvious signs is the large number of double-decker London Transport buses, still painted red, which come to Ceylon to die. Their corpses may frequently be seen by the wayside, mourned by passengers patiently waiting for a replacement.

Slightly older, and rather more interesting, ruins will be found in the sacred cities of Anuradhapura and Polonnaruwa, which were the capitals of the Sinhalese kingdoms for fifteen centuries, until about A.D. 1250. Dominating Anuradhapura are three great shrines, or dagobas, bell-shaped piles of solid masonry almost rivalling the pyramids in size. And here may also be found the famous 2,260-year-old Bo-Tree, reputed to be a sapling from the original tree under which the Buddha attained enlightenment.

The finest statues of the Buddha himself will be found at Polonnaruwa, which became the island's capital when Anuradhapura was abandoned (after repeated invasions from India) around A.D. 1100. A huge reclining figure, fifteen metres long, represents the Buddha at the moment of death; standing at its head is a smaller—but still impressive—statue with arms crossed and a look of downcast meditation or sadness. This is usually taken to be the Buddha's chief disciple Ananda, grieving over the departure of his master; but most modern scholars believe it to be another manifestation of the Enlightened One himself.

When Polonnaruwa was abandoned in turn, the Sinhalese kings retreated to the hill capital of Kandy, which still flourishes; during the Second World War, it was Lord Mountbatten's headquarters. Today its chief claim to fame is the Temple of the Tooth, from which every year caparisoned elephants accompanied by dancers, bands, and torch-bearers march in a spectacular procession honouring the sacred relic of the Buddha, which, however, never leaves the temple precincts.

Fifty kilometres south of Kandy is an equally holy spot, the 2,200-metre-high mountain, Adam's Peak—an equatorial Matterhorn, completely clothed with trees. Perched on its summit is a tiny temple, and every year thousands of pilgrims make the ascent up what must be the longest stairway in the world. The climb is not difficult, but can you imagine a stairway several kilometres long? I managed it once, and my legs were paralysed for the next three days. But it was worth it, for at dawn I saw the spectacle for which the peak is famous. As the sun rose, the perfectly triangular shadow of the mountain was cast on the clouds below, stretching for perhaps fifty kilometres into the west. It lasted for a magical ten or fifteen minutes, then faded out to cries of "Sadu, Sadu" ("Holy, Holy") from the pilgrims.

Of all Ceylon's archaeological wonders, however, the most remarkable—and certainly the most useful—is the enormous irrigation system which, for over two thousand years, has brought prosperity to the rice farmers in regions where it may not rain for six months at a time. Frequently ruined, abandoned, and rebuilt, this legacy of the ancient engineers is one of the island's most precious possessions. Some of its artificial lakes are ten or twenty kilometres in circumference, and abound with birds and wildlife.

Ceylon was once a hunter's paradise; Victorian "sportsmen" boasted of killing hundreds of elephants, bears, leopards, and birds beyond computation. There are now less than two thousand wild elephants in the island, most of them rigidly protected in great game sanctuaries. Wilpattu, the largest of these, is only a few hours' drive from Colombo, yet it seems so remote from man and all his ways that the visitor can easily imagine he is back in the primeval forest. No wonder Ceylon has become popular with makers of jungle-type movies; though Tarzan has not yet arrived, Elephant Bill and Mowgli are on the way, and who can forget Elizabeth Taylor facing the stampede of tuskers in the blazing finale of *Elephant Walk*? But the best film made in the country is still *The Bridge on the River Kwai*. If you want a vicarious trip to Ceylon, revisit David Lean's masterpiece next time it comes around.

The mountains, the jungles, the ruined cities, the lakes, the game reservations, the misty upland tea estates, will all have their advocates; but to me there is nothing that can compare with the beaches and the reefs. To see these at their best, you have to get well away from the towns. Drive south of Colombo for fifty kilometres, or fly across the island to the sun-drenched loneliness of the east coast, and you may find yourself the only human being on a crescent of soft white sand, enclosing a bay of the purest blue water, fringed by palm trees of astonishing height and slenderness.

The Indian Ocean is one of the last great unexplored regions of this planet. It is a curious experience to stand at the southernmost tip of Ceylon, Dondra Head, and to know that there is nothing but empty sea all the way south to the icy ramparts of Antarctica. After the swift sunset (for here you are only six hundred kilometres from the equator) the last lighthouse of the northern hemisphere will start to flash its warning above you, and far to the east you may see the flicker of its companion on the deadly Great Basses Reef. A long chain of barely submerged rocks, ten kilometres out at sea, this has been trapping ships ever since men started to sail the Indian Ocean. A few years ago my colleagues discovered the wreck of a large armed merchantman, carrying at least a ton of beautiful silver rupees. We nearly went bankrupt salvaging it, and I wouldn't wish sunken treasure on my worst enemy.

A better investment in time and effort would be hunting for emeralds and sapphires in the island's gem pits, famous since the days of Sinbad. His "Valley of Gems" was located in Ceylon, but the elephant-catching rocs which nested in the mountains above it are fortunately extinct.

Today's jumbos fly under their own power, and land at the very handsome Bandaranaike airport thirty kilometres north of Colombo. When you disembark here, you may encounter for the first time the phrase *"Ayu Bowan"* —which means both "Greetings" *and* "Goodbye."

With the adoption of a new constitution in 1972, Ceylon officially changed its name back to the ancient Sinhalese form. So—*"Ayu Bowan,* Ceylon; Sri Lanka, *Ayu Bowan."*

Either way, the country will still be just as beautiful.

In 1966, the island of Ceylon cut itself adrift from the rest of the world. It abandoned the seven-day week and reverted to the traditional Buddhist calendar, based on the phases of the moon. Thus each lunar quarter became a holiday (Poya Day), and the day before it a half-holiday.

The result was instant chaos. Only astronomers could make long-range appointments; it was useless planning to meet anyone on, for example, Monday week, since Monday (or any other day) might be Poya, and all offices and shops would be closed. This was merely an inconvenience as far as the country's internal affairs were concerned, but it caused utter confusion in all dealings with the outside world. Once every six weeks or so, as the moon drifted across the constellations, Poya Day fell on a Sunday and the Ceylonese were briefly in step with the rest of humanity. But most of the time, the tea-brokers of Mincing Lane could telephone their Colombo offices only three or four days out of every week, and they were never quite sure *which* days those would be. . . .

This attempt to put the calendar back a couple of thousand years was a move by Prime Minister Dudley Senanayake's mildly left-of-centre government to gain the approval of the priests, who have considerable influence in a country which is predominately Buddhist. But the opposition—led by Mrs. Bandaranaike, the world's first woman prime minister—was just as keen on the idea. Four years later, when her United Front party was back in power, it issued special stamps celebrating the anniversary of the Poya calendar and its hoped-for return to traditional virtues.

Despite its cost to the country, and the grumblings of the businessmen, no politician dared to attack this exercise in nostalgia; yet before it had completed its fifth year, it was quietly abolished and Monday, Tuesday, Wednesday . . . reentered the Ceylonese vocabulary. The sudden ending of the Poya calendar was a small but significant byproduct of a national tragedy which proved

that there is no way back into the past, and that the real
or imagined qualities of one's ancestors have little rele-
vance to unemployment, adverse trade balances, and the
other problems of modern life.

On the night of 4 April 1971, when the little island of
thirteen million people appeared reasonably contented
under a government which had been elected by a large
majority, a force of several thousand well-trained in-
surgents attempted to take over the country. Attacks were
launched upon many provincial police stations, which
were quickly overwhelmed; the rebels were thus able to
seize arms and ammunition and obtain temporary control in some rural areas.

Although there had been some earlier signs of trouble,
the government was taken almost completely by surprise.
It appealed for help—which promptly started to arrive
from a remarkable variety of sources. The countries
which rushed to the aid of Ceylon included (not necessarily in this order) the United Kingdom, the United
States, the U.S.S.R., India, Pakistan, the United Arab
Republic, Yugoslavia, and China. Ceylon is a land without
enemies; however, she frequently exasperates her friends.

In a few weeks of often bitter fighting, the insurrection
was put down and some fifteen thousand rebels and sus-
pects were rounded up in prison camps. To this day, the
organisers have never been clearly identified, though
a finger of suspicion pointed to the North Koreans. For
months, the local newspapers had been running paid,
closely printed essays of almost impenetrable turgidity,
describing the exploits of the heroic Kim Il Sun. Any
patient reader who *did* manage to penetrate the prose
automatically absorbed the principles of revolution, as
taught by an expert. Feeling that this was too much of a
coincidence, the government of Ceylon requested the
North Koreans to depart.

By the end of 1972, the country was back to normal,
though the human and economic cost of the tragedy would
never be fully assessed, and many of the factors that had

provoked it still remained. The ordinary tourist, jetting into Colombo's elegant Bandaranaike airport, will see no evidence of the island's recent traumatic experience. For though politicians and governments and social systems come and go, the land and the sun and the sea remain. There are few places on earth where they have combined in such exquisite proportions.

From the earliest times, Ceylon seems to have captured the hearts of travellers. Their reactions are preserved in the glamorous names they gave it: Serendip, Taprobane, the Resplendent Isle, Land Without Sorrow.

And to the English, of course, Ceylon has a special place —not merely because it was a colony for a century and a half, but because for most of that time it was a principal source of the tea without which the United Kingdom would come to a screeching halt. Early in 1940 some genius at the Ministry of Food realised this, and all the tea chests in the London docks were dispersed throughout the land—just ahead of the blitz. I spent my last civilian summer battling with innumerable bills of lading, trying to find exactly where the stuff had gone. It was then that I came across, for the first time, the names of the great estates and tea-growing districts—Uva, Dimbulla, Nuwara Eliya, Bandarawela. Little did I know, as the Victorian melodramas were fond of saying, that sixteen years later these exotic places would be the background of my own everyday life.

Until a few months ago, I was unable to account for this strange attraction to a country I had never seen, and had scarcely thought about, until my late thirties. On the frequent occasions when I was asked why I liked Ceylon I had plenty of answers, but they were never really convincing. There always seemed to be something missing; what it was, I discovered only recently—and on the other side of the globe from Ceylon.

Of course, some of the island's charms were blatantly obvious. To anyone who had survived a score and a half of English winters, the idea of spending Christmas sunbathing under the palms, or swimming in warm water of the purest blue, was irresistable. The best way of making the transformation—for those who have the time—is by the sea route. I can still remember watching the feeble December sun sink into the smoke of the London docks as I sailed away from winter forever. Twelve hours by jet is too short a time in which to relish such a miracle.

There is another reason why I like Ceylon; it is the right size—slightly smaller than Ireland. There is no point which cannot be reached from any other in a day's driving, over roads that are usually adequate and are often excellent, which, however, is more than can be said of most of the drivers *and* their vehicles. I vividly recall a Ceylon Transport Board bus which once shed its entire transmission fifty metres ahead of my car. As a hundred kilograms of metal bounced closer and closer, I unselfishly prayed that the world of letters would not sustain a major loss. I happened to have Gore Vidal riding beside me.

Geography and climate do not make a country, though they determine what kind of a country it will be. There are islands in the Pacific perhaps more lovely and more temperate than Ceylon, but they have no culture, no sense of the past—nothing to engage the intellect. Ceylon offers far more than the empty, mindless beauty that lured Gauguin to destruction; it has twenty-five hundred years of *written* history, and the ruins of cities that were once among the greatest in the world.

Over the millennia, successive waves of invaders have poured into Ceylon from the north; they brought with them their language, their technology, their art and, above all, their religion. When Buddhism declined in the country of its birth, India, it took root in Ceylon; and here, so its adherents claim, it is still to be found in its purest form. Certainly it survived almost five centuries of persecution and indifference, under three successive foreign regimes. The Portuguese, Dutch, and British occupations each lasted about a hundred and fifty years, very nearly fitting the convenient dates 1500–1650, 1650–1800, 1800–1950.

The Portuguese came with sword and cross; the Dutch with ledger and lawbook; the British with roads and railways. Each invasion had shattering but not wholly destructive effects on the indigenous life styles, and much of Ceylon's fascination today arises from this extraordinary mixture of cultures. A glance at the Colombo telephone directory is a cross section of history, an unplanned experiment in genetics. There are seven pages for Perera and almost as many for De Silva—with a generous sprinkling of Fernando, De Soysa, Dias, Gomez. Then, turning the pages and the centuries, one comes to much smaller numbers of latecomers straight from Joseph Conrad: Van Langenberg, Nathanielsz, Brohier, Jonklaas, Keuneman, Heyn, Koelmeyer.

All these are set in a background of pure Sinhalese: Senanayake, Hettiarachchi, Bandaranaike, Kirtisinghe, Atukorale, Wijeratne, Goonewardene. The Sinhalese constitute over 70 percent of the population, and theirs is now the official language—to the considerable disadvantage of the other main racial group, the Tamils. Some of the Tamil names scattered throughout the directory are real tongue twisters: Amirthanayagam (who is typing this article), Gnaasambanthan, Balasubramaniam, Duraippahpathar. And the commerce of centuries has brought the inevitable sprinkling of Mohameds and McLarens and Chettiars and Chens and Moosajees.

The population of the island now stands at about thirteen million, and most of them owe their lives to Dr. Hermann Muller. In case his name does not ring a bell with you, he was the man who discovered the useful properties of dichlorodiphenyltrichloroethane. And if that too is not instantly familiar—it is better known as DDT. Until the 1940's, malaria was endemic in Ceylon; a hundred thousand died of it in 1935. Despite a high birth rate, for centuries the population has been stabilised—

and enervated—by the mosquito. In what is now the text-book case of insect control by DDT, the carrier and the disease were virtually eliminated in the late 1940's.

As a result, the population doubled in thirty years, with the resultant problems of unemployment, inadequate social services, food shortages. A country so rich in nat-ural resources could support many more people than it does today, and at a much higher standard of living. But the *rate* of increase does not allow time for the necessary planning; the next generation has already arrived before the current one has prepared for it.

And it is hard to plan for the future, when the sun beats down from a cloudless sky, the waves whisper softly up the beach, the terraced fields of ripening paddy seduce the eye with their soft greens and golds. The men of the cold north believe that the tropics are hostile to civilisation because the struggle for existence can be too easily won. There is much truth in this, but there are also times when sun and drought can provoke as great a response as storm and snow. This happened, in Ceylon, before the beginning of the Christian Era, when a series of tremendous irrigation works transformed the island's dry zone into what must have been a fertile paradise. Some of the artificial lakes created then are many kilo-meters in circumference; there are thousands of these unfortunately named "tanks," linked by intricate net-works of canals. Only a stable, well-organised, and tech-nically advanced society could have undertaken such massive projects; such a society was seldom allowed to exist in peace for long, and successive invasions destroyed much of the work of the ancient engineers.

In modern times, many of the old tanks have been re-stored, and new ones have been built; these little inland seas, surrounded by mountains, are often so tranquil that they act as perfect mirrors for the passing clouds. Yet there can be few places on earth where the past seems more alive, more linked to the present by generations of toil and skill, all directed to the same end. For the modern electrically operated sluicegates may be set in time-worn stones that were carved before Caesar came to Britain. At the remoter tanks, the harassed survivors of Ceylon's once countless wild elephants may be seen coming to the water's edge to drink; in another decade, alas, they will be gone, though it is to be hoped a thousand or two will flourish indefinitely in the great game reservations where the only shooting allowed is by telephoto lens. Here, four or five hours' drive from Colombo's airport, is another world, even more ancient than that of the great engineer-kings. It is the world before man, when the only inhabitants of the island were deer and bear and leopard and boar and crocodile—all of which may be seen in an afternoon's Land-Rover ride.

The wildlife of the land has been famous for centuries, but only in this generation have we grown to know the wildlife of the sea—which is even more prolific, and far more colorful. Much of the island is surrounded by coral reefs, the new playgrounds of our age. It was these that first drew me to Ceylon and gave me the keenest moments of delight (and fear) I have ever known; only much later did I discover the lovely land that they guard against waves thundering across thousands of kilometres of open sea.

Ceylon is the last outpost of the Northern Hemisphere, jutting into the still largely unexplored emptiness of the Indian Ocean; it may be the base for the great undersea expeditions of the century to come. And the southern coast of the island, I realise now, is the source of the magic that holds me here.

I was born by the sea, and spent much of my childhood on the beach of a great curving bay in the west of England; no wonder that I have always been haunted by Housman's lines:

> Smooth between sea and land
> Is laid the level sand
> And here through summer days
> The seed of Adam plays.

But I lost the sea when I was about ten years old, and thereafter saw it only on holidays and brief visits. School, the civil service, the war, college, and a new career separated me further and further from the games of childhood. Even a year on the great Barrier Reef did not unlock the doors of memory. Not until I came to Ceylon did I fall in love with an exquisite arc of beach on the island's south coast and decide to establish a home there.

It takes a long time to see the obvious, and in this case perhaps there was some excuse. After all, there was little apparent similarity between dull gray English sea and turquoise Indian Ocean; between boardinghouses, Butlin camp, railway station—and an unbroken wall of closely packed coconut palms.

One day, after a lecture somewhere in the American Midwest, a young lady asked me just *why* I liked Ceylon. I was about to switch on the sound track I had played a hundred times before, when suddenly I saw those two beaches, both so far away. Do not ask me why it happened then; but in that moment of double vision, I knew the truth.

The drab, chill northern beach on which I had so often shivered through an English summer was merely the pale reflection of an ultimate and long-unsuspected beauty. Like the three princes of Serendip, I had found far more than I was seeking—in Serendip itself.

Ten thousand kilometres from the place where I was born, I had come home.

14

Willy and Chesley

One of the most beautiful—and, I suspect, influential—books ever written about space travel was published by Viking Press in 1949. It was called *The Conquest of Space* and consisted of paintings of astronomical scenes by Chesley Bonestell, with text by Willy Ley.

Willy had been a space enthusiast from the 1920's, and was one of the founders of the German rocket society, the *Verein für Raumschiffahrt*, which also numbered among its members a seventeen-year-old named Wernher von Braun. Though Willy was trained as a zoologist, it would be hard to set a limit to his interests, for he wrote hundreds of articles on almost every branch of science. His best-known work, *Rockets*, appeared in 1944, and its various revised and expanded editions still provide an invaluable history of the subject.

Chesley Bonestell was originally trained as an architectural draftsman, a fact that contributes to the often startling realism of his work. In the 1940's he began to produce his famous views of the moon, Saturn, and Jupiter, which attracted great attention when they were published in *Life* magazine. To millions, they were a foretaste of the coming age of space exploration, and even now it is fascinating to compare the reality of the Apollo photographs with Chesley's careful *pre*constructions.

After talking and writing about the possibility of space travel for half a century (in the early days, often to incredulous or downright derisive audiences) Willy died just a month before the first man reached the moon. The

131

reaction of his many friends was one of fury as well as grief; I give my own in the memorial essay which follows. And I am more than happy to say that the suggestion in its final paragraph was heeded.

My very last meeting with Willy Ley could hardly have occurred in more appropriate surroundings. I was descending the stairs of a New York subway when I spotted his figure some considerable distance ahead of me and was about to hail him when I had a sudden qualm: suppose it's really someone else and I make a fool of myself? Then I remembered the clincher: this was the Fortysecond Street and Fifth Avenue entrance, and the New York Public Library was immediately overhead. Anyone around here who looked like Willy *would* be Willy—and so it turned out.

Willy was more widely read, and owned a larger collection of books, than anyone I have ever met. His initial and lifelong interest was zoology—itself a sufficiently enormous subject—but the early European speculations about the possibility of space flight diverted him into rocket research and astronomy, as he recounted in his classic *Rockets, Missiles and Men in Space* (1968). Eventually, he became—to use the title of another of his works—equally at home *On Earth and in the Sky*, though he was certainly best known for his writing and lecturing on astronautics.

Nowadays, it is easy to forget that it required considerable courage to preach the possibility of space travel in the 1930's. In those days there were many experts (not a few of them now happily riding the space bandwagon) ready and eager to prove that all thoughts of escaping from earth were scientific nonsense. Willy Ley helped to educate the generation that turned the fantasy into hardware.

Although his most important work was undoubtedly *Rockets, Missiles and Men in Space* (which began in 1944 simply as *Rockets*), perhaps the book which was of the greatest inspirational value and aesthetic appeal was *The Conquest of Space*. In this handsome volume Willy matched his text to Chesley Bonestell's beautiful paintings of the moon and planets. When *The Conquest of Space* appeared in 1949, few of those who gazed entranced at the vistas of lunar landscapes, or the earth as seen from space, could have imagined that within a mere twenty years the vision of the scientist would have matched that of the artist. However, Chesley Bonestell's masterpiece—the crescent Saturn in the sky of its giant moon Titan—remains a spectacle which human eyes may not witness until the twenty-first century.

Yet Willy sometimes backed the wrong horse. In the first edition of *Rockets*, he implied that the German "secret weapon" was *not* a rocket, arguing that the performance required for long ranges demanded liquid propellants, which would not be practical for military applications. He was unlucky in his timing, for *Rockets* and the V.2 arrived at about the same moment. My colleague A. V. Cleaver, then director and general manager of the Rolls-Royce Rocket Division, has recorded meeting Willy in New York and being "astonished to find that, for some reason, he had decided that the rumours were a lot of nonsense. He spent much time and effort assuring me that his ex-countrymen were most unlikely to have developed such a weapon. . . . I argued weakly against these conclusions, but being very conscious of war-time security . . . forbore to tell him that I had personally heard the 'rumours' arriving" (*Spaceflight*, November 1969). But Willy's point about the military disadvantages of liquid-propelled rockets was well taken. As soon as possible, they were dropped in favor of solids; the Atlases and Titans have now been replaced by Minutemen and Poseidons.

Though Willy was an expert in many fields, and in

both appearance and accent was the typical German scholar, he was no pedant, and his countless articles were not only entertaining but frequently witty. Few people could demolish a crank with a more dextrous rapier thrust; perhaps the fact that in early days he must often have been taken for a crank himself gave him the technique for dealing with the genuine variety.

His writing was not, however, entirely restricted to fact; he was also the author of two interesting short stories. One—"Fog"—is an account of the general confusion during a political revolution, based on his own experiences in Germany. Better known is "At the Perihelion" (published under the name Robert Willey), which, as might be expected, is a soundly based tale of space flight. It introduced the idea of generating centrifugal "gravity" by setting two halves of a spaceship spinning at the opposite ends of a long cable. This technique was first tested on the Gemini 12 mission in November 1966.

Only once did I catch Willy out in a matter of scientific fact. In the preface to one edition of *Rockets*, he chided me for deserting the Northern Hemisphere to go to live in Ceylon. Having been under exactly the same misapprehension myself at one time, I gleefully told him to have another look at the map.

As countless aspiring writers and inquisitive science fans can testify, Willy was a kind-hearted and helpful person. During the Second World War, he earned my eternal gratitude by keeping me supplied with otherwise unobtainable issues of the American science-fiction magazines—particularly the handsome, large-size *Astounding Stories* of the lamented golden age.

I was on my way to the Apollo 11 launch when I heard the news of Willy's sudden death by heart attack. My immediate reaction—like that of all his friends—was one not only of sorrow but of something approaching anger. For forty-five years Willy had devoted the greater part of his life to the conquest of space. Although he had seen the total vindication of all his ideas, he had missed the final triumph by just four weeks. I could not help thinking of Dylan Thomas's famous lines:

> Do not go gentle into that good night;
> Rage, rage against the dying of the light.

Many have expressed the hope that one of the newly discovered lunar craters will be named after Willy Ley, when the committee of the International Astronomical Union charged with this task gets to work on Farside. But in any event, his books will ensure that his name is not forgotten. I know that I shall be consulting *Rockets* for the rest of my life; and whenever I do, I shall remember Willy.

When I pored over the Jovian and Saturnian vistas depicted in *The Conquest of Space,* back in those distant days when no rocket had ascended much more than a hundred kilometres from Earth, I never dreamed that the time would come when I myself would be collaborating with the artist. Yet in 1951 I did pay a somewhat novel tribute to Chesley Bonestell, when I wrote a story called "Jupiter Five" (in *Reach for Tomorrow*). This involved an expedition by "*Life* Interplanetary" to Saturn and Jupiter for the specific purpose of photographing the reality and comparing it with Chesley's century-old paintings.

Incredibly, just twenty years later, I was involved with Chesley on a project that would do precisely this. In 1971 NASA was about to launch the space probes Pioneer 10 and 11, which, if all went well, would radio back our first views of the mightiest of all planets. *Beyond Jupiter: The Worlds of Tomorrow,* our joint description of the project in illustrations and text, was beautifully produced by Little, Brown in 1972, and dedicated "To Willy, who is now on the Moon." It is quite uncanny to compare Chesley's painting of Jupiter from close quarters with the marvellous colour images later radioed back from the Pioneers, which are still heading on towards the stars, sending back information to the giant radio telescopes which follow their progress into the abyss.

They carry, as everyone knows, the famous plaques designed by Carl Sagan and his associates—the Space Age equivalents of those messages corked in a bottle and dropped into the sea. I was not aware of this (it was, in fact, a last-minute modification) when I wrote the concluding paragraphs of *Beyond Jupiter*, describing the ultimate fate of our first voyagers to the stars:

> As our space-faring powers develop, we may overtake them with the vehicles of a later age and bring them back to our museums, as relics of the early days before men ventured beyond Mars. And if we do not find them, others may.
>
> We should therefore build them well, for one day they may be the only evidence that the human race ever existed. All the works of man on his own world are ephemeral, seen from the viewpoint of geological time. The winds and rains which have destroyed mountains will make short work of the pyramids, those recent experiments in immortality. The most enduring monuments we have yet created stand on the Moon, or circle the Sun; but even these will not last forever.
>
> For when the Sun dies, it will not end with a whimper. In its final paroxysm, it will melt the inner planets to slag, and set the frozen outer giants erupting in geysers wider than the continents of Earth. Nothing will be left, on or even near the world where he was born, of man and his works.
>
> But hundreds—thousands—of light-years outward from Earth, some of the most exquisite masterpieces of his hand and brain will still be drifting down the corridors of stars. The energies that powered them will have been dead for eons, and no trace will re-

main of the patterns of logic that once pulsed through the crystal labyrinths of their minds.

Yet they will still be recognizable, while the Universe endures, as the work of beings who wondered about it long ago, and sought to fathom its secrets.

15

Mars and the Mind of Man

On 12 November 1971 the space-probe Mariner 9 arrived at Mars, went into orbit, and commenced taking a series of photographs which over the course of the next few months revolutionised our knowledge of the planet. The control center for the probe was at the Jet Propulsion Laboratory, Pasadena, and the day before the encounter a panel discussion was arranged a few miles away at Cal Tech. The moderator was the distinguished science editor of *The New York Times*, Walter Sullivan, and the panelists were Ray Bradbury, Carl Sagan, Bruce Murray (now director of J.P.L.), and myself.

At that time, of course, we did not know whether Mariner 9 would succeed or fail; indeed, we did not know until several weeks later, for ironically (though luckily for the meteorologists, who were able to study a unique phenomenon) Mars was completely shrouded by an enormous dust storm. But when the dust settled, an awesome terrain was revealed—volcanoes twice as high as Everest, and canyons that would stretch the width of the United States.

Our 12 November discussion—one of the most vigorous and entertaining I've ever enjoyed—was recorded and videotaped; when, a year later, we had an opportunity to examine the results of the mission, we all sat down and wrote rather more carefully considered afterthoughts. The resulting volume, *Mars and the Mind of Man* (New York: Harper & Row, 1973), with its selection of the superb Mariner 9 photographs, is a record of one of the truly great advances in our knowledge of the planets.

The passage that follows contains my unrehearsed and largely unprepared remarks on arrival at Cal Tech. The "afterthoughts" (I hope somewhat more literate) give my carefully considered views a year later.

I want to go along with Ray Bradbury's views on the importance of Edgar Rice Burroughs. It was Burroughs who turned me on, and I think he is a much underrated writer. The man who can create the best-known character in the whole of fiction should not be taken too lightly! Of course, there's not much left of his Mars, and his science was always rather dubious. I can still remember even as a boy feeling there was something a little peculiar about cliffs of solid gold, studded with gems. I think it might be an interesting exercise for a geology student to see how that phenomenon could be brought about.

Another writer I'd like to pay tribute to, partly because he lived such a tragically short time, was Stanley G. Weinbaum, whose *Martian Odyssey* came out around 1935. And then, of course, the other great influence on me was our Boston brahmin. Whatever we can say about Percival Lowell's[1] observational abilities, we can't deny his propagandistic power, and I think he deserves credit at least for keeping the idea of planetary astronomy alive and active during a period when perhaps it might have been neglected. He certainly did a lot of harm in some

---

1. Percival Lowell (1855–1916) founded his famous observatory at Flagstaff, Arizona, and brought the so-called canal controversy to the boil in the early part of this century. He claimed that Mars was covered with a spider's web of fine lines, looking very much like an airline map of Earth, which he believed to represent a vast irrigation system. (See also my comments at the end of this chapter.)

ways, but I think perhaps in the long run the benefits may be greater.

Anyway, I was very moved the other day when I visited the Lowell Observatory for the first time and actually looked through his twenty-six-inch telescope. He's buried right beside it; his tomb is in the shape of the observatory itself. I was distressed to find that his papers had been rather neglected and scattered around. As a result, I have initiated a series of events which may now result in his papers being classified and edited. Whatever nonsense he wrote, I hope that one day we will name something on Mars after him, and I'm sure that he won't be forgotten in this area.

And then, of course, you mentioned H. G. Wells. He certainly did a lot for Mars, and is still doing so, as you heard the other day. I don't know if movie director George Pal is here in the audience, but he has also done a lot for Mars, not to mention Los Angeles, in *The War of the Worlds*. He gleefully destroyed City Hall and a few other places around here.

We are now in a very interesting historic moment with regard to Mars. I'm not going to make any definite predictions, because it would be very foolish to go out on a limb, but whatever happens, whatever discoveries are made in the next few days or weeks or months, the frontier of our knowledge is moving inevitably outwards.

It has already embraced the moon. We still have a great deal to learn about the moon and there will be many surprises even there, I'm sure. But the frontier is moving on and our viewpoint is changing with it. We're discovering, and this is a big surprise, that the moon, and I believe Mars, and parts of Mercury and especially space itself, are essentially benign environments—to our technology, not necessarily to organic life. Certainly benign as compared with the Antarctic or the oceanic abyss, where we have already been. This is an idea which the public still hasn't got yet, but it's a fact.

I think the biological frontier may very well move past

Mars out to Jupiter, which I think is where the action is. Carl Sagan has just gone on record as saying that Jupiter may be a more hospitable home for life than any other place, *including Earth itself*. It would be very exciting if this turns out to be true.

I will end by making one prediction. Whether or not there is life on Mars now, there *will* be by the end of this century.

AFTERTHOUGHTS (1973)

Reading the transcript of this discussion is a curious experience, because it already seems to belong to another age—the prehistory of Martian studies. All of us knew, that November evening while Mariner 9 approached its moment of destiny, how important this mission might be, but I doubt if any of us would have dared to predict the full extent of its success. True, Mariner's cameras revealed no Martians carrying banners with the strange device BRADBURY WAS RIGHT (or even rival groups with NO—CLARKE WAS RIGHT). But what they did show was exciting enough. At last, we are zeroing in on the *real* Mars.

For most of this century, Mars has been haunted by the ghost of Percival Lowell, the man with the tessellated eyeballs. Mariners 4, 6, and 7 started to exorcise that ghost; Mariner 9 completed the job. The famous "canals" are gone forever. Why they ever appeared in the first place could be material for a valuable study of psychology and physiological optics—and, incidentally, the time is now more than ripe for a modern biography of Lowell, surely one of the most fascinating characters in the history of astronomy.

Now that we have good quality photographs of Mars, someone should compare Lowell's drawings with the reality to try to find just what happened up there at Flagstaff at the turn of the century. How was it possible for a man to sustain a self-consistent and extremely detailed

optical illusion (if that *is* what it was) over a period of
more than twenty years? How did he convince others of
his vision? What correlation, if any, was there between
the ability of other astronomers to see the canals and
their position on the Lowell Observatory payroll? These
are just a few of the questions that might be asked.

Recent work on the nature of vision has shown that
the eye is capable of feats which, a priori, one would have
said were completely impossible. "Eidetic imagery" is an
example which may be very relevant here. Dr. Bela Julesz
of Bell Labs discusses a case[1] where a subject was able
to store an *apparently random* pattern of ten thousand
picture elements—a 100×100 matrix of dots—and fuse
it *twenty-four hours later* with another array to produce
a stereoscopic image![2] As Dr. Julesz remarks with considerable understatement: "These experiments appear incredible," but they do suggest that the eye-brain system
has an astonishing capacity to store detailed images.
Could Lowell have built up, over a period of years, a
largely mental picture of Mars from the fleeting patterns
glimpsed through his telescope? The mind has an extraordinary ability to "see" things that are hoped for, assembling any chance visual clues that may come to hand—to
use a slightly mixed metaphor. When you are expecting to
meet a friend in a crowd, how often you see him before he
really appears!

If Lowell's Mars was indeed a largely subjective one, it
also had to be dynamic. It must have changed continually
with rotation, distance, seasons, to match the changing
appearance of the real Mars. Certainly a fantastic feat of
creative imagination, of the greatest interest to psycholo-
gists.

And although I'm speculating in areas with which I'm
completely unfamiliar, I'd like to stick my neck out just

---

2. *Foundations of Cyclopean Perception* (Chicago: University of Chicago Press, 1971).

a little further. Could there be some connection between Lowell's superbly maintained and brilliantly proselytised delusion (remember, plenty of other observers "saw" the canals) and a similar phenomenon of our own time? I don't think there can now be any doubt that hundreds of intelligent, sober, and altogether reliable citizens have honestly "seen" brightly moving lights in the sky and all the other familiar UFO phenomena.[3] How many of these originated in some such process as Lowell's canals?

However, back to the real Mars. It now appears that, by one of those ironies not uncommon in science, the earlier Mariner results caused the pendulum to swing too far to the other extreme—away from the hopelessly romantic view of Mars. For the few years from 1965 to 1972 Mars was a cosmic fossil like the moon—no, not even a fossil, because it could never have known life. The depressing image of a cratered, desiccated wilderness was about as far removed from the Lowell-Burroughs fantasy as it was possible to get.

There were some, undoubtedly, who accepted the new "revelation" with considerable relief—even glee. Now there would be no further fear of that dreaded cry in the night, "The Martians are coming! The Martians are coming!" We were comfortably alone in the solar system, if not the universe.

Well, perhaps we are, but it seems more and more unlikely. The new Mars that has suddenly emerged from the Mariner 9 photos, a world of immense canyons and volcanoes and erosion patterns—and, dare one say, dried-up sea-beds?—is a much more active and exciting place than we would have ventured to hope, only a few years ago. Lowell and Company may yet have been partly right, for the wrong reasons.

It is not really a coincidence that, while Mariner 9 was being built, the first positive evidence for the chemical

---

3. Anyone who still doubts this should read Allen Hynek's *The UFO Experience* (Chicago: Henry Regnery, 1972).

evolution of complex organic molecules beyond the earth was being discovered. The basic building blocks of life were being found in *meteorites*, of all places, perhaps as hostile an environment as could be imagined. In view of this, and the obvious signs of past water activity shown in the Mariner photos, the biologists will have some explaining to do—if there is *no* life on Mars.

Meanwhile, we science-fiction writers had best be cautious for a few years—perhaps until U.S. or U.S.S.R. soft-landers start to do some detailed reporting in the mid-seventies. For myself, I'm already a little embarrassed to see that *The Sands of Mars* (1951) contains the sentence *"There are no mountains on Mars"* (in italics). Well, it took over twenty years to shoot *that* one down, so it had a good run for its money. And on the plus side, we now have some perfectly beautiful photos of Martian sand dunes, so at least my title was completely valid. The sands of Mars have survived very much better than the oceans of Venus. (Poor Venus—what a hatchet job the Mariners and Veneras did on her! But that's another story.)

There are some not-very-bright and/or badly educated people who complain, with apparent sincerity, that scientific research destroys the wonder and magic of nature. One can imagine the indignant reaction of such poets as Tennyson or Shelley to this nonsense, and surely it is better to know the truth than to dabble in delusions, however charming they may be. Almost invariably, the truth turns out to be far more strange and wonderful than the wildest fantasy. The great J. B. S. Haldane put it very well when he said: "The universe is not only queerer than we imagine—it is queerer than we *can* imagine."

I feel sure that Mariner 9—and its successors—will provide many further proofs of this statement. We have already learned an instructive lesson from the moon, which is becoming more complicated and more interesting with every expedition. The same thing will happen with Mars. Whether we find life or not, we will discover things which

we could *never* have imagined. And these will provide material for the deeper and richer fantasies of the future,
just as the earlier observations inspired the fantasies of
the past.

And the beauty of it is—we can have it both ways!
When men are actually living on Mars, at the turn of the
century, they will be reading the latest works of the lucky
science-fiction writers who are starting their careers now,
at the beginning of the fourth golden age. Yet at the same
time, they will still be able to enjoy, from their new per-
spective, the best of Wells and Burroughs.

And, I hope, of Bradbury and Clarke.

16

The Snows of Olympus

Of all the discoveries made by Mariner 9, one of the most impressive was an enormous volcano, twice as high as Everest and five hundred kilometres across the base. It was located at a spot which the old map-makers, peering at the tiny telescopic image of Mars, had noted as being of unusual brilliance, and had given the astonishingly prescient name *Nix Olympica*, the Snows of Olympus. Now that the real nature of this marking is known, the name has been changed to *Mons Olympica*, Mount Olympus. Yet it can hardly be doubted that there *is* snow there from time to time, in that thin and freezing atmosphere so far above the Martian plains. However, it will be snow of carbon dioxide, not water.

Early in 1973 *Playboy* magazine asked me to write a short essay on any subject I pleased. Here is the result, which I consider one of my best pieces of nonfiction. (It was also, possibly, a fond farewell to the magazine which had published most of my nonfiction *and* shorter fiction for almost twenty years, because soon afterwards—for the reasons given in Chapter 18—I decided to concentrate entirely on novels. But, from time to time, I will undoubtedly change my mind.)

In 1972, for only the third time in history, mankind discovered a new world. It happened first in 1492. The impact of that discovery was immediate, its ultimate benefits incalculable. It created a new civilisation and revivified an old one. The second date, not quite so famous, is

147

1610. In the spring of that year, Galileo turned his primitive "optic tube" towards the moon and saw with his own eyes that Earth was not unique. Floating out there a quarter of a million miles away was another world of mountains and valleys and great shining plains—empty, virginal—awaiting, like Michelangelo's *Adam*, the touch of life. And 362 years later, life came, riding on a pillar of fire.

With the end of the Apollo program, there will now be a short pause until much cheaper transportation systems are developed. Then we will return and the history of the moon will begin. But it may be quickly overshadowed by a greater drama, on a far more impressive stage. The third new world was not found by sailing ship or by telescope; yet, like the two earlier discoveries, it was a shocking surprise that resulted in the overthrow of long-cherished beliefs. No one knew that such a place existed, and when the evidence started to accumulate, early in 1972, many scientists were literally unable to believe their eyes.

This new world has nearly twice the diameter of the moon and is almost four times as large as both Americas. And it has the most spectacular scenery yet discovered anywhere in the universe. Think of the Grand Canyon, the greatest natural wonder of the United States. Then quadruple its depth and multiply its width five times, to an incredible seventy-five miles. Finally, imagine its spanning the whole continent, from Los Angeles to New York. Such is the scale of the canyon that is carved along its equator.

Yet even this is not the planet's most awesome feature, for it is dominated by volcanoes that dwarf any on earth. The mightiest, *Nix Olympica*—The Snows of Olympus— is almost three times the height of Everest and more than three hundred miles across. Those volcanoes are slumbering now, but not long ago they were blasting into the thin atmosphere all the chemicals of life, including water: there are dried-up riverbeds that give clear indication of recent flash floods—the first evidence ever found for running water outside our earth. It even appears that this may be a young world, geologically speaking; if life has not already begun there, that will be yet another surprise.

By now, you may have guessed the identity of this new world. It is Mars—the *real* Mars, not the imaginary one in which we believed until Mariner 9 swept aside the illusions of decades. It will be years before we absorb all the lessons of this, the most successful robot space mission ever flown; but already it seems that Mars, not the moon, will be our main order of extraterrestrial business in the century to come.

This news may be received with less than enthusiasm at the very moment when NASA's budget is being cut to the bone and voices everywhere are calling for an attack on the evils and injustices of our own world. But Columbus did more for Europe by sailing westward than whole generations of men who stayed behind. True, we must rebuild our cities and our societies and bind up the wounds we have inflicted upon Mother Earth. But to do this, we will need all the marvelous new tools of space—the weather and communications and resources satellites that are about to transform the economy of mankind. Even with their aid, it will be a difficult and often discouraging task, with little glamour to fire the imagination.

Yet "where there is no vision, the people perish." Men need the mystery and romance of new horizons almost as badly as they need food and shelter. In the difficult years ahead, we should remember that the Snows of Olympus lie silent beneath the stars, waiting for our grandchildren.

17

Introducing Isaac Asimov

Shortly after "The Snows of Olympus" had appeared in *Playboy* (December 1974), my dear friend Isaac Asimov also wrote an essay on *Nix Olympica* for his regular series of science articles in *The Magazine of Fantasy and Science Fiction*. These have been appearing every *month* for more than twenty years; sometimes, when Isaac flags, they are merely brilliant. You can guess what title he had chosen for the piece; and you can imagine his rage when he picked up the December *Playboy*. So, as he duly reported, he changed the name of *his* essay to something entirely different: "The Olympian Snows."

In June 1974 Isaac made his first visit to the United Kingdom—a somewhat belated one because, like Ray Bradbury and Stanley Kubrick, he absolutely refuses to set foot in an airplane. (One day I hope NASA tempts him with a ticket for the space shuttle.) The purpose of the trip was to address the British section of Mensa, the the 150-plus IQ organization, and as by a fortunate coincidence I happened to be in the country at the time, I was asked to make the introduction.

This was too good an opportunity to be missed, and I spent several days composing an impromptu collection of carefully contrived insults, duly delivered in the Commonwealth Hall, London, on 14 June. Isaac had no warning; nevertheless, as you can judge for yourself, *his* riposte gave as good as he received. The whole encounter, leading into a fine lecture by Isaac on the value and importance of science fiction, is available from British Mensa,

Limited, 13, George Street, Wolverhampton, WV2 4DF, England.

Still, I have the last word—at the moment. *My* entry in the *Encyclopaedia Britannica* is longer than Isaac's.

Well, Isaac—I've lost my bet. There *are* more than five people here. . . .

Ladies, gentlemen—and in case there are any robots or extraterrestrials present—gentlebeings. . . .

I'm not going to waste any time *introducing* Isaac Asimov. That would be as pointless as introducing the equator, which, indeed, he's coming to resemble more and more closely.

I'd like to begin by claiming a small part of the credit, if that's the word, for getting him here, since I feel that I helped to start him travelling. A couple of years ago I was involved in a project to take the *Queen Elizabeth 2* to the Apollo 17 launch. As it happened I couldn't make it myself, and it was another ship anyway, but the press gang got Isaac on board, together with Bob Heinlein and Norman Mailer and lots of other characters, and everyone had a wonderful time. That started him making like Marco Polo, and he didn't even look scared when I met him on the *Canberra* last year on his way to the June 30 eclipse. Indeed, he was busy holding forth on the pleasures of travel—to Neil Armstrong.

But we still haven't been able to get him into one of these new-fangled flying machines. Try it, Isaac, you'll like it. They show some very interesting movies on the big jets. Last time I flew the Atlantic, they screened *A Night to Remember*. They won't show you *that* when you sail next weekend on the *QE2*. And talking of *A Night to Remember*, I've persuaded the man who produced it, my good friend Bill MacQuitty, to come along this evening. Afterwards, he would like to give you some hints about useful things to do in the lifeboats.

Incidentally, Isaac, your sedentary habits have deprived you of an honour you richly deserve. For a long time I've been embarrassed—and I don't embarrass easily —by the fact that you haven't got the Kalinga Prize for science writing, because no one has ever done more to earn it. After I'd spent several years lobbying for you, my Indian friends finally said to me: "What's the use? The prize has to be accepted in New Delhi, and everyone knows we'll never get the *meshugana* to come here." (For those of you who aren't Sanscrit scholars, a meshugana is someone whose DNA has been acquired in an irregular manner; it's a close relative of a *yenta*. And that's a private joke between Isaac and Harlan Ellison.)

Let's hope this problem will soon be solved. In a few months the Suez Canal will be open, so you'll be able to get to India easily. And then of course you must come down to Ceylon and I'll take you on an underwater safari from my diving base at Sharkhaven and introduce you to my favorite scorpion fish and my pet octopus.

Of course, I should hate anything to happen to Isaac.... The rumour that there is a certain rivalry between us should have been put to rest, once and for all, in my recent *Report on Planet Three*. For those of you foolish enough not to have obtained that small masterpiece, the dedication reads: IN ACCORDANCE WITH THE TERMS OF THE CLARKE-ASIMOV TREATY, THE SECOND-BEST SCIENCE WRITER DEDICATES THIS BOOK TO THE SECOND-BEST SCIENCE-FICTION WRITER.

On the whole I've kept my side of the treaty—though sometimes I have to confess that there's a better man than either of us sitting up there behind his hippie-proof chain fence in Santa Cruz. Do you realise, Isaac, that *at this very minute*, while we're wasting time here, Bob Heinlein is racing away at a thousand words an hour?

What a man. I've just had a four-thousand-word letter from him, finished at 3 A.M. after he'd collected my latest Nebula Award (I had to get that in somewhere). Most of it was about his wartime engineering adventures with you, Sprague de Camp, Will Jenkins, and Ron Hubbard. Incidentally, whatever happened to Ron? He was a damn good writer. He could easily make ten cents a word nowadays.

Still, Isaac, I'm sure you'll soon make up for lost time, when you get back to your four electric typewriters. Isaac is the only man who can type separate books simultaneously with his two feet as well as his two hands. One day the literary scholars will be writing theses, trying to decide which book was typed with which foot.

It seems only yesterday that *Opus 100* came out, and now he's past the halfway mark of his *second* century. And I'm proud to say that I can reveal some top-secret information—his future writing plans. My private Plumber's Unit—you can hire ex-White House staff pretty cheaply these days—did a little job on Isaac's apartment while he and Janet were whooping it up on the *France*. So here are a few of the forthcoming titles: "Asimov's Guide to Cricket"; "The Asimov Cookbook"—to be followed immediately by "The Asimov Exercise Book"; "The Asimov Limerick Collection" (remind me to tell you, Isaac, what sent the Countess of Clunes to the Cloister); "Asimov's Kama Sutra"; "Asimov's Guide to Watergate"; "1001 Nights with Isaac Asimov"; "Asimov's Hints on Survival at Sea."

These are just a few, and as I went through my microfilms I discovered what his game plan is. Well, Isaac, I'm much more ambitious than you. I don't want to write three hundred books in sixty years. *I* intend to write sixty books in three hundred years. . . .

It's an awesome output—and it's not true, as some have suggested, that Isaac is actually a robot himself. If you want proof, ask any of the thirty young ladies at the

Globe last Wednesday what they had to do to get his autograph.

After all this flattery, though, I must regretfully issue a few words of criticism. Have you ever thought of the entire forests this man has destroyed for woodpulp? He's an ecological catastrophe. The other day I asked my computer to work out the acreage he's devastated. Here's the answer.

"I know I've made some poor decisions lately, but I'm feeling much better now" . . . ooops, sorry.

"Why don't you take a stress pill . . . ?"

Ah, here we are: "Deforestation by Asimov—5.7 times ten to the sixteenth microhectares." Well, now you know. All those beautiful trees, turned into Asimov books.

Of *course* I'm not jealous. But I must admit to a teeny touch of envy when I heard all about his autographing sessions at various bookstores. The only time I tried that, they got the date wrong. I had planned to have some sandwich-board men walking up and down outside with UNFAIR TO CLARKE on the front and ASIMOV GO HOME on the back. But the scheme fell through because, despite all my careful instructions, they spelt his name correctly. So I've taken other modest remedial steps, with the help of the IRA. Isaac, you made a big mistake when you took up residence at the Oliver Cromwell Hotel.

Finally—a delicate but unavoidable matter which I will try to handle with my usual exquisite tact. As you probably know, Isaac was born in Russia. By a curious error, all the references give a date which makes him younger than I am, which is obviously ridiculous. Now, had he gone to one of the excellent language schools there, he might by now be speaking perfect English. However, he emigrated to Brooklyn.

So even if you don't understand a word he says, *please* don't embarrass our distinguished guest. When he's finished, forget your British reserve and give him a round of hearty applause . . . pitter, pitter, pitter.

Gentlebeings—there is only one Isaac Asimov. Here is that Isaac Asimov.

THE ASIMOV RESPONSE

You will not be surprised to know, ladies and gentlemen, that I have made a study of introductions. I have had plenty of opportunity for it, and I have discovered two laws: (1) A *dull* introduction is better than a *clever* one. It is easier to follow a dull introduction. People are delighted to hear you by contrast. (2) The second law is that a *short* introduction is better than a *long* one. A long one wears you out. Therefore, the *very worst* kind of introduction is a long, clever introduction—and Arthur *knows* this!

Let me tell you the kind of guy Arthur is. When he met me on the *Canberra* and he saw that I was perfectly at ease and had overcome my fear of travelling and was standing there with nothing between myself and the sea but some thin steel, he said, "Isaac, at great expense I have persuaded the captain of this ship to show *The Poseidon Adventure*."

I'll tell you what kind of a guy Arthur C. Clarke is. He receives letters from people saying: "Enclosed you will find a very complicated and deep theory of mine explaining the entire universe," and out fall seventy-five thin sheets of onionskin, scribbled all over with indecipherable writing, interspersed by ridiculous diagrams. He sends it all back and he says: "Dear Sir, It is clear to me that your theory is of the greatest interest, and it disturbs me that I cannot give it the attention it deserves. However, by a curious coincidence, my friend Isaac Asimov, who lives at the Oliver Cromwell Hotel, etc., etc., is *particularly* interested in just this sort of thing"—and you should see my mail!

And I will tell you right now, that from here on in, I won't mention him at all. Let us instead talk about science fiction, which, after all, is what we both do—I because I am a great writer, and Arthur because he is a stubborn writer.

18

Life in Space

During the course of the last thirty years, my records indicate that I have published approximately five hundred articles and short stories, which is a very modest output compared with that of many writers, especially those who must depend upon authorship for their entire living. But in 1973, with the appearance of *Rendezvous with Rama*, I decided to do no more short pieces and to write novels— or *nothing*.

There was no single reason for this decision, which I was reluctant to adopt. The inevitable loss of energy with increasing years was certainly a factor; equally important was the feeling that I should write nothing that anyone else could do as well, or even half as well. This at once cut out all the science journalism that had occupied so much of my younger days.

I could leave this important field with a clear conscience, for now there are plenty of good science writers— though not enough jobs for them. (Another reason for me to make way for a younger generation.) The Clarke-Asimov Treaty, for which I refer you to the dedication of *Report on Planet Three*,[1] is not entirely a joke.

Moreover, in the fast-moving field of space exploration, any article dated so quickly that its half-life was only a few years. True, in this respect I have been lucky with my nonfiction, some of which has stayed in print much longer than I wished; I even had to apply euthanasia to

---

1. See Chapter 17, p. 153.

some of my first books on astronautics, when I found that they were being read by a generation unborn when they were written.

Yet all my *fiction* was still in print and appeared to be earning more with each new edition. It was also more fun (though more hard work) to write. And if *I* didn't write it, certainly no one else would. . . .

Financially and psychologically, therefore, the choice was a logical one. In 1973 I wrote what I fully intend to be my last straight science article, "Life Beyond the Earth," which appeared in the Time/Life Nature Science Annual for 1974.

Many years ago, according to a somewhat dubious legend, a distinguished astronomer received the message: "Is there life on Mars? Cable thousand words. Hearst." In return, the newspaper tycoon got his thousand words: "Nobody knows"—repeated five hundred times. Hearst may well have thought the scientist uncooperative. Yet he was certainly honest, and even today his reply is still valid, not only for Mars but for everywhere outside Earth. Nobody knows.

But the situation is changing rapidly. After years of frustration, during which we could do nothing except speculate on the basis of hopelessly inadequate evidence, we are now racing towards the answer. And perhaps many answers, because there are now tantalizing hints that the universe is not so hostile to life as was once supposed. Until recently, scientists believed that life could exist only in a narrow zone of space centered on the earth. Sunward, towards Venus, it was too hot; starward, beyond Mars, too cold. Today it seems that this attitude was unduly pessimistic, and the pendulum is swinging in the other direction. Very recently the biologists have turned their attention towards the giant planet Jupiter, orbiting in the frigid wastes beyond Mars. And in 1973, a case was made by Carl Sagan for still a more unlikely abode of life: Titan, the largest of Saturn's moons.

The discovery of life in outer space would be the greatest prize in the whole history of science. Yet before we can even begin the search, we must try to settle one important point. Just what do we mean by "life"—even the qualified version usually mentioned in discussions of this kind: "life as we know it?" It turns out that life has something in common with obscenity. Everyone can recognize it, but no one can define it. Almost all definitions break down somewhere along the line.

Living things grow; but so do the crystals of such inanimate stuff as ice and rubies. Living things have complex and usually symmetrical patterns; compare a starfish and a snowflake. They reproduce themselves; but the atomic particles called neutrons do a better job of that in a uranium-fission chain reaction. No—it is not so easy to tell when an aggregate of atoms and molecules enters the magic state of life, and we must beware of our earth-centered bias when we concoct a universal definition of life.

Even the frontier between the living and the nonliving on *this* planet cannot be clearly drawn. When we go down into the submicroscopic world of the viruses, we meet entities—better not call them organisms—which can be completely inert for unlimited periods, behaving and looking like crystals of inanimate matter. But in suitable circumstances—the presence of the right kind of food, such as may be provided by living cells—they spring into action and start to reproduce themselves. However, because they are not capable of both independent and active existence, we hesitate to consider viruses "living."

With the amazing developments of biochemistry and genetics that have taken place during the last quarter century, it has been recognized that there is indeed no sharp boundary between the living and the nonliving. Instead, we have a continuous gradation—or evolution—

from simple inanimate structures to complex living ones:
Atoms → simple organic molecules → proteins → self-
reproducing molecules (DNA, the hereditary material) →
single-celled organisms → simple plants → animals →
man.

The progression can be extended backward (before
atoms, there were subatomic particles) and doubtless
forward, beyond man. The earlier arrows represent trivial
increases in complexity, which can be duplicated in the
laboratory. However, the jump from self-reproducing
molecules to living cells involves a gigantic leap, for a cell
is a whole factory of biochemical processes, intricately
organized. If a protein may be somewhat fancifully compared to a spark plug or a crankshaft, a living cell is all
of General Motors.

Nevertheless, it is widely believed that this increase in
complexity is a natural and inevitable process, bound to
occur given enough time and suitable conditions. The
current theory is that it took place on the primeval earth
in warm seas or ponds under the influence of sunlight,
volcanic heat, and lightning. First there was chemical
evolution: the formation of dilute broths of carbon, hydrogen, nitrogen, and oxygen compounds, particularly those
known as amino acids. These are the constituents of protein molecules, the basic components of flesh and blood,
and hence of all living creatures. Later—we may never
know the full details with certainty—this chemical evolution led to biological evolution, that is, the sequence
plants-to-animals-to-man.

With this theory to guide us, we are now in a far better
position to search for life beyond the earth. And, rather
surprisingly, there is no need to leave our planet to begin
that quest. Nature is kind enough to provide us free
samples from outer space, in the form of meteorites that
fall to earth perhaps dozens of times a year. Most fall into
the ocean, and only a fraction of those that descend on
land are ever pinpointed accurately enough to be recovered. Although stories of rocks falling from the

heavens go back to the dawn of history, until the beginning of the nineteenth century most scientists regarded all such reports as nonsense. In 1807, for example, Thomas Jefferson remarked that he would prefer to believe that two Yankee professors had lied rather than accept that stones had fallen from the sky. Once it was proved that we did have genuine visitors from space, great efforts were made to collect and study them. And, naturally, scientists examined meteorites to see if they showed any traces of life or its byproducts. Some researchers claimed to have found such evidence— even including fossils! But all of these early reports turned out to be mistaken; it was not realised that any meteorite, after it has lain on the ground or been on display in a museum, will have picked up some of the chemical signatures of life. If it was touched just once by sweaty hands, that could be enough to contaminate it.

By a fortunate combination of circumstances, a giant leap in meteor chemistry, helping to set the stage for the current excitement over extraterrestrial life, occurred soon after Neil Armstrong's first step onto the moon. It was the result of luck—plus planning. For several years, NASA scientists had been preparing to receive the first samples from the moon. They had devised new techniques of chemical analysis, far more sensitive than any that were known before, to extract the maximum amount of information from a few grams of rock more precious than crown jewels. (If the Apollo program had ended with the first flight, the cost of the specimens would have been about a billion dollars a kilogram.)

Only two months after the Apollo 11 splashdown, NASA's cosmochemists received an unexpected free bonus. On September 28, 1969, a shower of metorites fell near the Australian town of Murchison, forty kilometres north of Melbourne, and almost 100 kilograms of fragments were collected. One small stone was rushed to NASA's Exobiology Division at Ames Research Center in California, where it was examined by a team headed by

Dr. Cyril Ponnamperuma. Thus, for the first time, a meteorite that had not lain around long enough to acquire earthly contamination could be investigated by techniques capable of detecting the faintest traces of life.

The results were of the greatest scientific importance. The Murchison meteorite contained substantial quantities of organic compounds, including six of the amino acids that occur in natural protein—as well as twelve that do *not* occur in any proteins known on Earth.

This discovery did not demonstrate the existence of life elsewhere. But it did prove that some of its basic ingredients, of quite a high degree of complexity, occur in a most unpromising environment. Somewhere out in space the atoms of carbon, hydrogen, oxygen, and nitrogen managed to get together to create several of the letters in the alphabet of life. They also created a few in an unknown script—which, in some other part of the universe, may build up words, then sentences, then the complex libraries that are living creatures. We can only speculate, but our speculations can now be based on some firm facts.

It does seem surprising that small lumps of rock drifting between the planets should contain such clear evidence of the early stages of chemical evolution; outside the earth, amino acids and similar compounds would be expected to be formed only in the same environments believed to have fostered them here: oceans or atmospheres of large planets. Perhaps this was indeed the case with the chemicals in the Murchison meteorite, for it may be that meteorites are the debris of a planet that was destroyed in some cosmic catastrophe. However, in the last few years evidence has accumulated that planets are not necessary for the initial stages of chemical evolution; it can proceed very nicely in so-called empty space. This discovery is even more unexpected than the finding of organic compounds in meteorites. Until our generation, space was regarded as a perfect vacuum—the total absence of all matter. We now know that this is far from being so.

True, space is a vacuum as far as human life is concerned; the tragic deaths of the three Soyuz 11 astronauts were proof of that. The air we breathe at sea level contains some twenty million million million molecules of oxygen and nitrogen per cubic centimetre; the atmosphere 150 kilometres up, where the closest satellites orbit, contains only a few billion. And in deep space between the stars, the figure is down to about ten atoms—mostly hydrogen—per cubic centimetre.

Such amounts of material may seem too small to be of any practical importance, but the sheer volume of space is so enormous that even at a few atoms a thimbleful, the gas drifting through it could outweigh all the stars and constitute most of the matter in the universe. Until recently, however, it was impossible to study—or even to detect—the interstellar gas clouds, because they are quite invisible to ordinary telescopes. The spectacular rise of radio astronomy has changed all that. The molecules in most simple gases have natural rates of rotation that give rise to extremely high frequency radio waves. Hydrogen, for example, has a characteristic frequency of 1,420 megahertz (a wavelength of twenty-one centimetres). This distinctive, sharply tuned signal from space was first detected as long ago as 1951, practically in the Old Stone Age of radio astronomy. It has since become one of the most valuable tools in mapping the distribution of material in our galaxy and uncovering the pattern of its spiral arms.

As the astronomers searched the skies with more and more elaborate radio equipment, they discovered the radio signatures of other gases, with the result that, during the 1960's, a wholly new and quite unexpected science was founded—interstellar chemistry. By 1973 more than a score of compounds had been discovered floating between the stars: almost all of them essential to the processes of life. Most important was water, discovered in 1968; then in rapid succession came ammonia, formaldehyde, methyl alcohol, formic acid, and more complicated substances like methyl acetylene. It may only be a matter of time before someone turns up the first amino acids, the precursors of life.

This proliferation of organic compounds in such unlikely locations as meteorites and the depths of near-empty space certainly suggests that almost everywhere Nature is setting the cosmic stage for the advent of life. The final proof of this, however, still eludes us. We might have found it on the moon, though no one really hoped to discover living organisms on our satellite as it is today. But there was at least a faint possibility that at some time in the past the moon might have had a temporary atmosphere, running water, and transient seas, in which life could have evolved. As it turned out, Apollo samples contained only microscopic traces of organic compounds; there is not the slightest hint that life existed on the moon before July 20, 1969.

Nevertheless, we cannot certify the moon as completely sterile; it is only 99.9 percent probable. So far, only a dozen men and a handful of robots have scratched around in a few square kilometres on a world about the size of Africa. The first few metres of surface material have been baked by an unshielded sun and churned by meteorites for billions of years; any organic materials, if they ever existed, would long ago have been blasted into space. In some shielded region deep inside the moon, there may yet be surprises waiting for future explorers. However, it would be unwise to bet much money on it.

Meanwhile, the search for life has gone farther afield to the other planets. For a long time it seemed that Venus was the most likely prospect. In size it is almost a twin of Earth, and although it is a good deal closer to the sun, it was rather confidently hoped that its unbroken ceiling of clouds would reflect much of the solar heat back into space, so that the temperature on the surface would be reasonable.

Alas, radio observations, together with the U.S. Mariner and U.S.S.R. Venera space probes, totally destroyed the dream of a subtropical Venus. The surface, mapped by radar in 1973, is as pocked and torn as the moon's. And the atmosphere (almost entirely carbon dioxide) is so dense that the pressure is a hundred times that at the earth's surface—about the same as a kilometre *underwater* on our planet. Such pressure in itself is no handicap to life, but unfortunately the surface temperature is around 900° F. (500° C.). This is a consequence of the "greenhouse effect," painfully familiar to anyone who has ever left a closed car parked in direct sunlight for a few hours. Glass acts as a heat trap; so do many gases, especially the carbon dioxide in our atmosphere and that of Venus. There is far more carbon dioxide on Venus, and the planet is much nearer to the sun. So we had better seek cooler climes on Mars, which has long been the center of biological interest in the solar system, and still is, despite several changes in fashion.

Until the Space Age, all our knowledge of Mars was derived from observations made by astronomers straining their eyes to make out faint details at the very limits of vision. If you look at the moon through a sheet of dirty glass, you will have a fair idea of a good telescope image of Mars. The problem is compounded not only by distance (Mars is never closer than sixty million kilometres) but by the distorting effects of the earth's perpetually trembling atmosphere. And to make matters worse, most of the Martian features—except the sparkling polar caps—are lacking in contrast, so they would be hard to observe from Earth even if viewing conditions were perfect.

It is scarcely an exaggeration to say that despite more than a hundred years of devoted effort, all our knowledge of Mars before 1972 was hopelessly misleading. But in that year the fabulously successful Mariner 9 mapped the planet from pole to pole with high-definition TV cameras and measured characteristics of its surface and atmosphere. The first great surprise was that Mars is not, as had been previously thought, a geologically dead world. It is dominated by huge volcanoes—up to twenty-five thousand metres high—which appear to have been recently active. And, most exciting of all, there is evidence of dried-up riverbeds, which may have been in full flood only a few seconds ago, as the cosmic clock measures time. This first sign of the existence of liquid water outside Earth, like the discovery of organic molecules in meteors and in space, goes counter to all preconceived ideas. Today, there can be no liquid on Mars. The pressure in the atmosphere is too low (less than a hundredth of its value here) for water to exist except in the form of ice or vapor, both of which have been detected. However, conditions may have been different in the past and perhaps were favorable for the evolution of life. If life was ever able to obtain a foothold on Mars, it could have adapted itself to the rugged conditions that now prevail.

The futility of speculating about hypothetical Martian life forms is all too obvious when we consider the incredible variety of organisms that our planet has brought forth. However, Martian life forms would have to meet certain fundamental conditions which would make them seem very strange to us. First of all, they could not breathe; the very tenuous atmosphere contains no free oxygen. This would seem to rule out all animals, or equivalent active life forms, and leave only plants and bacteria. But nature is endlessly ingenious; we can imagine parasitic, slow-moving organisms subsisting on Martian vegetation. If such creatures exist—and it is unlikely—they would need tough, impermeable skins to meet the problems of conserving water and resisting the extreme cold (perhaps as low as a couple of hundred degrees Fahrenheit below, in winter nights).

This sort of argument can go on forever, and will undoubtedly still be raging when the first Earth robot meets the first Martian. For, almost certainly, we will have to get down to the actual surface of the planet before we can learn anything about its indigenous life; observations from space can never be conclusive, as has been proved by satellite studies of Earth. Even though we *know* what we are looking for on our home ground, it is extremely difficult to prove, by photographs from space, that our planet is inhabited. The seasonal color variations might be due to chemical reactions or dust deposited by winds—as indeed appears to be the case for at least some of the changes observed on Mars. And, ironically, the most conspicuous productions of life on this planet could never be identified by distant observers; for example, who could imagine that Australia's two-thousand-kilometre-long Great Barrier Reef was built by minute coral animals?

We should not have to wait long for a close-up look at Mars. NASA plans to launch the Viking probe to Mars in 1975; it will carry TV cameras and soil-sampling devices which, with a great deal of luck, may prove the existence of biological reactions on the planet. But it is probable that Russian scientists will answer the question first; they have missed no opportunity to send spacecraft to Mars and they succeeded in making a safe landing there with Mars 3 in December 1971. Unfortunately—and one can imagine the feelings of the scientists concerned—the TV camera sent back only a single blank picture before transmission ceased. No one knows what went wrong; there was no indication of a malfunction. Perhaps the months-long storm that was raging toppled the spacecraft. It is rather less likely that a Martian sneaked up on it from behind with a pair of wire cutters.

Quite recently—and, again, this is one more unexpected turn of events—the specialists on space life, exobiologists, have focussed their attention beyond Mars into the even colder and more remote realm of the giant planets. The closest (never less than six hundred million kilometres distant) and largest (eleven times the diameter of the earth) is Jupiter, which has long been known to have a deep atmosphere of hydrogen and helium, plus traces of methane and ammonia, both of which are involved in the processes of life. However, because of Jupiter's great distance from the sun, it was believed to be so cold that liquid water could not possibly exist there, and even the ammonia might be in the form of frozen crystals. This picture was based on the theory that Jupiter received heat only from the sun and had no internal sources of warmth. Measurements of its infrared radiation now show this idea to be false; down beneath those impenetrable clouds, Jupiter has some home fires burning. Not literally, of course; although its atmosphere consists largely of combustible gases, there cannot be a fire where there is no free oxygen. Jupiter's heat may be caused by radioactivity, or perhaps by the planet's slow contraction under its immense gravitational field.

Whatever the explanation, there may be levels in the Jovian atmosphere as warm as Earth—and loaded with huge quantities of organic compounds. If some form of life has not started to evolve there, it will be both a surprise and a disappointment. For the atmosphere of Jupiter is believed to be similar to that of Earth when the first green scum of algae started to gather on the shores and ponds four billion years ago.

Some of Jupiter's twelve moons, too, possess at least one of the crucial preconditions for life: the presence of water. Using a telescope that was equipped to measure infrared reflectivity, a team of scientists found that water frost covers much of the surface of two of the satellites, Europa and Ganymede. Lesser amounts of frozen water seem to exist on two other moons.

The close-up study of Jupiter and its environs began with space probe Pioneer 10's fly-by in December 1973. No space probe has yet ventured past Jupiter to Saturn, strangest and most beautiful of planets with its halo of encircling rings. But very recently some curious news has come from one of the many moons of this eerie world.

Saturn has ten known satellites, and there are probably many others still undiscovered. The largest is Titan, big enough to be regarded as a planet if size were the only criterion; 4,500 kilometres in diameter, it is almost a twin of the planet Mercury. For thirty years Titan has had the unique distinction of being the only satellite known to have an atmosphere; methane was found there in 1944. However, methane gas is transparent—and Titan appears to be covered with reddish clouds. Moreover, measurements of its infrared radiation, which should provide a clue to its temperature, have given anomalous results. Like Jupiter, Titan is much hotter than was expected, and the theory has been advanced that it gets its warmth the way Earth and Venus do, from the greenhouse effect of its atmosphere. This idea has recently been taken further by Carl Sagan.

Sagan, director of Cornell University's Laboratory for Planetary Studies, wears at least three hats: astronomer, biologist, and general scientific gadfly. Long an enthusiastic hunter of extraterrestrial life, he believes that Titan's atmosphere must be chiefly hydrogen to produce the thermal blanketing effect that is observed. But this explanation poses other problems; hydrogen, the lightest of all gases, would quickly escape from so small a world. It must therefore be continually replenished, presumably from Titan's interior, by some mechanism comparable to the volcanoes and geysers on Earth. Thus there may be local hot spots on Titan, where the temperature is high enough for water to exist in liquid form. If this admittedly long chain of conjecture is correct, there may be areas of Titan favorable for the evolution of life.

The reddish color of Titan's clouds, together with the amount of methane known to be present in its atmosphere, suggests that some kind of chemical evolution is in progress there. Jupiter's famous Great Red Spot may contain similar materials; NASA's Cyril Ponnamperuma, in his book *The Origins of Life,* describes the manner in which electric sparks, passed through a mixture of methane and ammonia in his lab, produced translucent, orange-red compounds. Many of the substances involved in the processes of life have this characteristic color—blood is an obvious example—but this may be mere coincidence.

Even if Titan has some hot spots, most of the satellite is probably far colder than any region on Earth. Perhaps the very best that one can hope for is minus 100° C., and this still seems optimistic. Nevertheless, if Sagan's theory of local volcanic activity is correct, life might have evolved in warm gases and then moved out into the frigid wilderness. There, under a pale pink sun, it would feed on the immense quantities of organic material littering the somber Titanian landscape. If it achieved intelligence, its religions would doubtless give praise to a beneficent creator for having provided so hospitable a world. Titan may indeed be a nice place to live in; but we wouldn't want to visit—not in person at least. However, our robot probes may pay a call in the early eighties, and perhaps they will signal back the news that Earth is not unique.

Only the biologists, however, will be really excited by the discovery of a few microbes or mosses on another world. To most people, extraterrestrial life will not be important unless it can talk back to us, preferably in a friendly way.

In all its long history, our own planet has brought forth only one intelligent species out of the millions that have existed. Is this proportion a fair sampling for the rest of the universe? We simply do not know. Some scientists argue that even if life is commonplace among the stars, intelligence may be extremely rare, possibly even unique to Earth. That is a lonesome thought; better the hostile aliens of the old pulp magazines. And the idea that our planet has been visited by alien beings from space is an old one. Early in this century, an eccentric writer named Charles Fort spent years collecting news items and historical reports that suggested such events, which he published in a maddeningly discursive book called *Lo!* Later writers have rehashed Fort's material, but such purely historical evidence, though intriguing and suggestive, can never give proof. Nothing much short of an alien skeleton or an artifact of a superior technical civilization could provide that. No one has ever found any such evidence of space visitors, perhaps because archaeologists have never seriously looked for it. Now some of them may be doing so; it is unlikely that they will be successful, but the prize is so enormous that it is worth a small investment in time and effort.

A much more promising approach is to search for radio signals from space—artificial signals, as opposed to the quintillions of watts of natural radio emissions from the stars. On a small scale in the United States, and a much larger one in the U.S.S.R., this search has already begun. But so far the equipment used is inadequate. Recently a study was carried out for NASA—Project Cyclops—to estimate what would be required for a really determined attack on the problem. A vast array of radio-telescopes several kilometres in diameter would greatly increase the probability of success. However, the cost would amount to billions of dollars spread over many years.

Meanwhile, it has been suggested that we have already received signals from space, without knowing it. During the late 1920's, a series of curious radio echoes was picked up by European short-wave listeners. Echoes of Morse code transmissions from a station at Eindhoven, Holland, were detected after delays of up to twenty seconds. If they were genuine echoes from some reflecting surface, this delay would imply that the echoes had travelled almost six million kilometres.

This phenomenon has never been satisfactorily explained. In 1973 a young British amateur astronomer named Duncan Lunan advanced the sensational theory that the Eindhoven echoes could be an attempt by a visiting space probe to signal its presence. By an ingenious analysis of the time delays involved, Lunan claimed that he had deciphered messages from the hypothetical probe, including the statement that it originates from the direction of the constellation Bootes.

The argument is too tenuous to carry much conviction, and is somewhat reminiscent of the logic that once led amateur cryptographers to "prove" that Bacon wrote Shakespeare. With a little imagination, almost any reasonably long sequence of letters or numbers can yield any message desired. Using Lunan's own arguments, and a little misplaced ingenuity, it may be possible to prove that his space probe's place of origin was not Bootes but Brooklyn.

Despite disappointments and false alarms, the search will continue; surely, someday, it will succeed. It is anyone's guess whether success will come tomorrow—or a thousand years from now. Perhaps there is only one thing of which we can be sure. The universe is infinitely surprising, and our knowledge still very limited. As the past history of science has so often shown, one cannot outguess nature. The most unlikely possibility of all is that we are completely alone in this galaxy of a hundred billion suns, this cosmos of a hundred billion galaxies. It is far more probable that the universe is crawling (also walking, swimming, flying, rolling, hopping) with life. And it will come in three varieties:

Life as we know it.

Life as we don't know it, but can imagine it.

Life as we can't possibly imagine it.

Which will we meet first?

"Nobody knows. . . ."

19

Last(?) Words on UFO's

My reason for not writing any more articles applied, a thousand times more powerfully, to book reviews. In any case, I had refused to do reviewing for years because I knew perfectly well that most of the books sent to me would be by personal friends, and I did not wish to risk turning them into enemies. This does not deter professional reviewers, but they either are protected by a cloak of anonymity or have skins of rhinoceros hide.

But all rules must have occasional exceptions, and when the *New York Times Book Review* asked if I would look at two of the very few serious books on UFO's, I rather nervously accepted the assignment. I had studied the subject for at least a decade *before* the unfortunate phrase "flying saucer" came into circulation, and wanted to see if there was anything further to add to my essay "Things in the Sky" *(Report on Planet Three)*.

Herewith the result, including the footnote that I asked the *Times* to print. I am happy to say that the *Times* kept its side of the bargain, and I never received a single letter.

And if this reprinting triggers any correspondence, it will not be read. After more than forty years, starting with Charles Fort's *Lo!* (1931), I am no longer interested in any further books, or letters, *about* UFO's.

But I am still interested in UFO's themselves.
Mildly.

UFO's Explained. By Philip J. Klass. Illustrated. New York: Random House, 1974.

The UFO Controversy in America. By David Michael Jacobs. Foreword by J. Allen Hynek. Illustrated. Bloomington & London: Indiana University Press, 1975.

One beautiful evening in the spring of 1964, Stanley Kubrick and I, after weeks of brainstorming, finally decided that we'd make a film about space. In a mood of cheerful exhilaration, we walked out onto Stanley's penthouse balcony to admire the view over Manhattan.

Suddenly, we noticed a brilliant star rising in the south. As it climbed steadily up the sky, I assumed it was the Echo balloon satellite—but to our astonishment, when it reached the zenith it apparently came to rest, hovering vertically above the city.

We rushed indoors to collect Stanley's Questar. By the time we had set it up, the object was moving again, and we followed it for almost ten minutes before it disappeared over the northern horizon. But even through the telescope, it remained a featureless point of light.

Somewhat shaken, and clutching at the one remaining straw, we checked the daily listing of Echo transits in the *Times*. The satellite was not due for another hour. . . .

Of the half-dozen UFO's I have seen (the first in the 1930's, long before anyone heard of "flying saucers") this is the only one that ever made me lose any sleep, and reading these books brought it vividly back to mind. For no one who has not been through such an experience can really appreciate how a combination of unusual circumstances, inaccurate information, and heightened emotional states can completely delude observers who consider themselves intelligent and level-headed.

Would you believe that the flaming fragments of a meteor or a reentering satellite *a hundred miles away* would be mistaken by honest witnesses for a nearby vehicle with distinct rows of lighted windows (rectangular, at that!)? Philip Klass proves conclusively that this has happened not once, but several times. Will respectable, law-abiding citizens, often holding public office, fabricate photos and evidence? Same answer. . . . No unbiassed reader of *UFO's Explained* can doubt that, even when there is some real, external stimulus, much of the event reported lies in the mind of the beholder.

Yet the book's title is misleading. Though Mr. Klass provides a welcome breath of sanity in a field where it is sadly lacking, he only explains *some* UFO's. From the multitudinous nature of the beasts, there will always be some that forever defy explanation, often because the truth is too ludicrous to be accepted. (Yes, Virginia, some UFO's *are* swamp gas. Some are owls. Some are bits of paper. Some are spiders' webs. And who would have believed a burning golf ball, releasing the explosive energy of its core as it bounded through the night?)

No wonder that the poor U.S. Air Force was shot out of the sky when it went into battle with the UFO's, all Xerox machines firing. Dr. Jacobs's scholarly study makes it abundantly clear that there was never any high-level "conspiracy" to hide the "facts" of extraterrestrial visitations, though one can hardly blame those who thought so at the time. The Air Force's pitiable attempts to sweep the whole annoying mess under the carpet, culminating in the hilarious shambles of the "Condon Report," make pathetic reading.

So does the coldly factual demolition of the fakers and psychopaths who have bedevilled the subject; Dr. Jacobs is content to let their own words condemn them. Lest we forget, thousands of nitwits believed George Adamski in 1955 when he "flew" to the moon and saw lakes, rivers, and cities on the far side—now, of course, revealed in all its barren detail by American and Soviet cameras. But Adamski's successors are still in business; it's a pity there's no consumer-protection law giving, say, a thousandfold refund to purchasers of their demonstrably fraudulent books. Yet perhaps the various flying saucer cults provide a harmless hobby for unstable personalities. (I'm not referring, of course, to the many *serious* researchers who think that UFO's are important, and whose work is made so difficult by the crackpots.)

These two books are partly complementary, and belong in any library of the ten (to be generous) volumes worth reading on the subject of UFO's. Klass is a complete skeptic; Jacobs thinks that there may be a hard core of phenomena still unexplained by contemporary science. (This reviewer inclines to the same belief on Mondays, Wednesdays, and Fridays.) One theory that can no longer be taken very seriously is that UFO's are interstellar spaceships. If there are as *few* as a million of these roaming round our galaxy, I shall be very much surprised; but when they do turn up, we'll know in sixty seconds. They won't hang around for centuries, looking for a place to park.

Oh, I'm glad you asked. Our UFO turned out to be Echo, after all. For some extraordinary reason, the *Times* hadn't seen fit to print this splendid transit in its daily ephemeris, though two others were given that evening and simple arithmetic would have disclosed the third. The illusion that the object was hovering at the zenith had, like so many UFO's, multiple causes: (1) We were too excited to observe calmly, (2) it is almost impossible to judge the angular movement of anything vertically overhead, (3) bright moonlight had obliterated the star background which normally betrays satellite motion.

The whole incident once again underlines the only lesson we have so far learned from UFO's. They tell us absolutely nothing about intelligence elsewhere; but they do prove how rare it is on Earth.

Arthur Clarke, the science-fiction writer (*2001: A Space Odyssey,* and so on), is a resident of Sri Lanka. He believes that UFO's need a few decades of benign neglect, and threatens to sue *The Times* if it forwards any correspondence relating to this review.

20

When the Twerms Came

In Chapter 18 I explained the reasons why I decided to write only novels, and no more nonfiction. But what about *short* fiction?

The short story has been the mainstay of science fiction, and perhaps its best and most characteristic literary form. Today, there is practically no market for "mainstream" short stories, but there is such an endless flood of science-fiction anthologies that it's becoming virtually impossible to think of new titles for them. The editors are being reduced to dates and numbers: *Orbit 55, The Best Science Fiction Stories of 1984.* (Now *that* should be an interesting collection.)

And the short story—in all lengths up to the twenty-thousand-word *novella*—has always been my own favorite. This is not entirely a matter of laziness, though a job which can be completed in a few days or weeks, rather than perhaps years, has much to commend it. From the nature of the beast, you can get closer to literary perfection in a short story than in a novel. Thus, of my own two best works, I suspect that, in an absolute sense (if that has any meaning), *Transit of Earth* is "better" than *Imperial Earth.*

By 1972 all the short stories I had ever written, with the exception of a few pieces of juvenilia, had been published in six collections, concluding with *The Wind from the Sun.* But I have one story left, which didn't make the deadline for that volume. *Playboy* wanted to turn it into a psychedelic cartoon, and I can still recall the anxiety

with which the editors nervously revealed the layout to
me. They were so relieved when I laughed.

When the Twerms Came is all of four hundred words
long, and unless I write another seventy thousand words
of short fiction—which seems most unlikely—it will have
no place to call its own. So here, for better or worse, is my
last short story.

Goodbye, Rudyard. Nice knowing you, Ernest. Thanks
for everything, H. G. . . .

We now know (little consolation though this provides)
that the Twerms were fleeing from their hereditary
enemies the Mucoids when they first detected Earth on
their far-ranging Omphalmoscopes. Thereafter, they re-
acted with astonishing speed and cunning.

In a few weeks of radio-monitoring, they accumulated
billions of words of electroprint from the satellite News-
pad services. Miraculous linguists, they swiftly mastered
the main terrestrial languages; more than that, they
analysed our culture, our technology, our political-
economic systems—our defences. Their keen intellects,
goaded by desperation, took only months to identify our
weak points, and to devise a diabolically effective plan of
campaign.

They knew that the U.S. and the U.S.S.R. possessed
between them almost a teraton of warheads. The fifteen
other nuclear powers might only muster a few score giga-
tons, and limited deliver systems, but even this modest
contribution could be embarrassing to an invader. It was
therefore essential that the assault should be swift,
totally unexpected, and absolutely overwhelming. Perhaps
they did consider a direct attack on the Pentagon, the Red
Fort, the Kremlin, and the other centers of military
power. If so, they soon dismissed such naïve concepts.

With a subtlety which, after the event, we can now

ruefully appreciate, they selected our most compact, and most vulnerable, area of sensitivity. . . .

Their insultingly minuscule fleet attacked at 4 A.M. European time on a wet Sunday morning. The weapons they employed were the irresistible Psychedelic Ray, the Itching Beam (which turned staid burghers into instant nudists), the dreaded Diarrhea Bomb, and the debilitating Tumescent Aerosol Spray. The total human casualties were thirty-six, mostly through exhaustion or heart failure.

Their main force (three ships) attacked Zurich. One vessel each sufficed for Geneva, Basle, and Berne. They also sent what appears to have been a small tugboat to deal with Vaduz.

No armorplate could resist their laser-equipped robots. The scanning cameras they carried in their ventral palps could record a billion bits of information a second. Before breakfast time, they knew the owners of every numbered bank account in Switzerland.

Thereafter, apart from the dispatch of several thousand special delivery letters by first post Monday morning, the conquest of Earth was complete.

21

The Clarke Act

During my years of tax-imposed exile from Sri Lanka, when I was compelled to absent myself from the country for at least six months out of every twelve, I filled in some of the time very profitably and pleasantly by lecturing. Much has been written about the horrors of the American lecture circuit, but I was lucky. I had the best agent in the business (Bill Colston Leigh), and when I gave him the dates between which I was available, I could confidently leave all the details to him and his efficient staff. I would arrive in New York, be handed an itinerary and a wad of tickets—and thereafter merely followed instructions, like a well-programmed computer. Never once did I have to send the Leigh Bureau the equivalent of G. K. Chesterton's fabled telegram, which in my case might have read: AM IN GOPHER CROSSING SOUTH DAKOTA STOP. WHERE SHOULD I BE? And only once did I miss an engagement—when a metre of snow descended on my Greyhound bus *and* the lecture venue towards which I was heading, thus neatly cancelling out the audience as well.

But after twenty years, and hundreds of talks, the novelty had definitely worn off. I was getting tired of my own jokes, and was tempted to give an award for an original question from the floor. As I was getting paid as much for an hour's vocalisation as for my first *year* in the British civil service, I felt that I was taking money under false pretences. And, worst of all, it was no longer a challenge.

An attempt to price myself out of the market showed
no signs of succeeding when the government of Sri Lanka
came to the rescue. For years, I had been badgering its
ministers and ambassadors—in Colombo, London, Washington, and New York, in and out of season—concerning
the urgent necessity of a reciprocal tax treaty with the
United States. In 1974 my propaganda paid off.
Minister of Finance Dr. N. M. Perera (representative
of a rare and endangered species, the Trotskyites) earned
my eternal gratitude by setting up a scheme to "invite
foreigners of goodwill to reside in Sri Lanka" without
paying *any* local tax on money they brought into the
country. Moreover, they would be allowed to bring in car
and household goods free of customs duty. Almost the
only restrictions were the very reasonable ones that they
must not engage in any business or any political activity
which might affect the security of the state.

You will not be surprised to hear that I was the first
person to be issued one of the new "Resident Guest"
visas, dated 28 February 1975—or that the act setting
up the scheme is popularly known by my name.

So my annual six-month pilgrimage was over, and a
great weight was lifted from my mind. No longer was I
a literary Vanderdecken, forced to wander by 707 and
VC-10 across the face of the globe. I could begin to enjoy
a civilised and leisurely life, in the one place I really
wanted to be.

However, the end of lecture touring did not mean the
end of lecturing, or at least of public appearances. From
time to time there would be a request from so prestigious
an organisation, or in such an attractive locale, that I was
unable to say no. Thus, when I.B.M. invited me to spend
a week with several hundred of its executives, and my
old friends Betsy and Walter Cronkite, at the magnificent
Acapulco Princess Hotel, I was happy to accept; I can
understand why, a little later, Howard Hughes chose this
for his last earthly abode. (I feel I can also claim, fairly
confidently, to be the only man to have attended an

I.B.M. banquet wearing a sarong, which, *pace* Dorothy Lamour, is *male* attire in Sri Lanka). An engagement in the matchless Sydney Opera House was another irresistible attraction, especially as it was almost twenty years since I had been to Australia. I suspected there might have been some changes, and there were; about the only thing I could recognise was the bridge, and English was now spoken everywhere. I never heard a word of Strine,[1] though doubtless it still flourishes in the Outback.

Yet now, as I become more and more sedentary, even the prospect of giving a lecture in the Taj Mahal or the Kremlin is not, in itself, a sufficient incentive. There has to be some intellectual challenge as well, but when one is in the late fifties new ideas are difficult to come by, requiring hard work to dredge them up from the subconscious. A really new lecture involves weeks sweating blood in my office, with the sun shining gloriously outside and the waves breaking languidly over the reef a few kilometres away. . . .

My decision (some may consider it much overdue) only to speak when I have something new to say thus limits me to one, or at most two, engagements a year, preferably with a brief pause for shopping at Hong Kong or Singapore on the way home. The three pieces that follow are the first results of this policy, each involving an amount of labor which I will seldom be willing to undertake again. But for that very reason, I also hope that they will be as rewarding to the reader as, in the final delivery, they were enjoyable to me.

---

1. The indigenous Australian dialect, which sounds like Cockney and is (or was) almost as incomprehensible.

22

Technology and the Limits of Knowledge

In 1972 the Smithsonian Institution's National Museum of History and Technology arranged a series of lectures in honor of Frank Nelson Doubleday, to celebrate the seventy-fifth anniversary of the publishing firm he founded. The series was introduced by the museum's director, the distinguished historian Daniel Boorstin (now Librarian of Congress), and the five speakers in the 1972–1973 session were Saul Bellow, Daniel Bell, Edmundo O'Gorman, Sir Peter Medawar, and myself. The lectures were black-tie affairs, given in one of the Smithsonian's marble halls to a distinguished selection of the Washington elite.

The overall theme of the first session (there have since been two others) [1] was "Technology and the Frontiers of Knowledge," a theme interpreted pretty liberally by the speakers. Saul Bellow's "Literature in the Age of Technology" opened in fine style by quoting an "extraordinarily silly" statement from a book by one Arthur C. Clarke. I count it as a noble (and possibly unique) example of my self-restraint that, though I had the last word, I refrained from comment.

And I still do.

---

1. All published in *The Frontiers of Knowledge* (New York: Doubleday, 1975).

The twin subjects of this talk are *technology* and *knowledge,* and whenever I hear that second word I am reminded of a little poem popular at Oxford about a hundred
years ago:

> I am the Master of this College;
> What *I* don't know isn't knowledge.

This claim was, of course, immediately challenged by a
rival establishment:

> In all Infinity
> There is no one so wise
> As the Master of Trinity.

Unless my memory is betraying me yet again, the
modest first couplet emanated from Balliol and was attached to Benjamin Jowett, the theologian and Greek
scholar.

Today, of course, a man like Dr. Jowett lies squarely
on the far side of the famous culture gap. Most of today's
knowledge consists of things that he didn't know, and
couldn't possibly have known. This is not because of the
sheer increase in knowledge, though that has been enormous. But the very center of gravity of scholarship has
now moved so far that there are vast areas where any
high school dropout is better informed than the most
highly educated man of a hundred years ago.

Much of this change may be linked with the other
gentleman I mentioned just now: the Master of Trinity,
than whom, et cetera. The most famous holder of this post
was J. J. Thomson, discoverer of the electron; and *that*
discovery marks the great divide between our age and all
ages that have gone before. It transformed technology,
and it transformed knowledge.

The electronic revolution and the devices it has spawned
are now changing the face of our world and will determine
the very structure of future society. And the discovery of
the electron led, of course, directly to modern physics and
the picture of the universe we have today—so much more

complex and fantastic than could possibly have been imagined by any philosopher of the past.

One is almost tempted to argue that most *real* knowledge is a byproduct of technology, but of course this is an exaggeration. Much that we know about the world around us has been derived over the centuries by simple naked-eye observation: in some important fields, like botany and zoology, this is still partly true. Yet, even here, we could never have understood the facts of simple observation without the technology represented by the microscope and the chemical laboratory. It can be argued that we do not really know anything until we understand it; mere description is not enough. Ancient naturalists such as Aristotle and Pliny recorded many of the basic facts of genetics; it is only in our time that the secret of the DNA molecule was uncovered, after a gigantic research effort involving every weapon in the technological armory from computers to electron microscopes.

There are those who think that this is a pity, and who somehow feel that knowledge is "purer" in direct proportion to its lack of contamination with technology. This—literally!—mandarin attitude is a consequence, as J. D. Bernal has remarked, of "the breach between aristocratic theory and plebeian practice which had been opened with the beginning of class society in early civilization and which had limited the great intellectual capacity of the Greeks."[2]

This failure of the Greeks—and the Chinese—to fuse technology and knowledge in a truly creative manner is one of the great tragedies of human history; it lost us at least a thousand years. Both these great civilizations had plenty of technology, some of a very high order, as Joseph Needham and others have shown. Nor did the Greeks despise it, as is often imagined; the myth of Daedalus and the reality of Archimedes show their regard for sophisticated mechanics.

---

2. *Science in History* (London: Pelican, 1969), 2: 375-376.

Yet, somehow, these brilliant minds—of whom it has also been truly said that they invented all known forms of government and couldn't make one of them work—missed the breakthrough into experimental science; that had to wait for Galileo, two thousand years later. How near the Greeks came to the modern age you can see, if you have sufficient influence and persistence (it took me three visits and a letter from an admiral), in the basement of the National Museum of Athens. For there, tucked away in a cigar box, is one of the most astonishing archaeological discoveries of all time, the fragments of the astronomical computer found by sponge divers off the island of Antikythera in 1901. To quote from Dr. Derek Price:

> Consisting of a box with dials on the outside and a very complex assembly of gear wheels mounted within, it must have resembled a well-made *18th Century* [my italics] clock. . . . At least twenty gear wheels of the mechanism have been preserved, including a very sophisticated assembly of gears that were mounted eccentrically on a turntable and probably functioned as a sort of epicyclic or differential gear system.[3]

Looking at this extraordinary relic is a most disturbing experience. Few activities are more futile than the "What if . . ." type of speculation, yet the Antikythera mechanism positively compels such thinking. Though it is over two thousand years old, it represents a level which our technology did not reach until the eighteenth century. Unfortunately, this complex device described merely the planets' apparent movements; it did not help to *explain* them. With the far simpler tools of inclined planes, swinging pendulums, and falling weights, Galileo pointed

---

3. "An Ancient Greek Computer," *Scientific American,* June 1959. On the copy he sent to me, Derek Price has written hopefully, "Please find some more." I am afraid that the most advanced underwater artifact I have yet discovered is an early-nineteenth-century soda-water bottle.

the way to that understanding, and to the modern world. If the insight of the Greeks had matched their ingenuity, the industrial revolution might have begun a thousand years before Columbus. By this time we would not merely be pottering around on the moon; we would have reached the nearer stars.

One of the factors which has caused this gross mismatch between ability and achievement is what might be called intellectual cowardice. In the extreme case, this is best summed up by that beloved cliché from the old-time monster movies: "This knowledge was not meant for man." Cut to the horrified faces of the villagers, as the mad scientist's laboratory goes up in flames.

The noncelluloid version is a little less dramatic. It consists of assertions that something can never be known, or done, rather than that it *shouldn't* be. But often, I think, the underlying impulse is fear, even if the only danger is the demolition of a beloved theory. Let me give some examples which are relevant to the theme of this talk.

It's grossly unfair to judge anyone by a single piece of folly; few of us would survive such a critique. But I have never taken Hegel seriously—and have thus saved myself a great deal of trouble—because of the *Dissertation on the Orbits of the Planets,* which he published in 1801. In this essay, he attacked the project then under way to discover a new planet occupying the curious gap between Mars and Jupiter. It was philosophically impossible, he explained, for such a planet to exist. . . . By a delightful irony of fate, the first of the asteroids had *already* been discovered a few months before Hegel's unfortunate essay appeared. I do not know if he issued a revised edition, but Gauss remarked sarcastically that this paper, though insanity, was pure wisdom compared to those that Hegel wrote later.[4]

---

4. Willy Ley, *Watchers of the Skies* (New York: Viking, 1963), p. 320.

Some of my best friends are Germans, but I cannot resist quoting an even more splendid specimen of Teutonic myopia. When Daguerre announced his photographic process in 1839, it created such a sensation that some people simply refused to believe it. A Leipzig paper found that Daguerre's claims affronted both German science and God, in that order:

> The wish to capture evanescent reflections is not only impossible, as has been shown by thorough German investigation, but . . . the will to do so is blasphemy. God created man in his own image, and no man-made machine may fix the image of God. . . . One can straightway call the Frenchman Daguerre, who boasts of such unheard-of things, the fool of fools.[5]

One would like to know more about the subsequent career of this critic, whom I have resurrected not merely for comic relief but because he provides an excellent introduction to a much more instructive debacle. This time, I am happy to say, the culprit is a Frenchman; I would not like anyone to accuse me of nationalistic bias. He is the philosopher Auguste Comte. In the second book of his *Course of Positive Philosophy* (1835), Comte defined once and for all the limits of astronomical knowledge. This is what he said about the heavenly bodies:

> We see how we may determine their forms, their distances, their bulk, their motions, but we can never know anything of their chemical or mineralogical structure; and much less, that of organized beings living on their surface. We must keep carefully apart the idea of the solar system and that of the universe, and be always assured that our only true interest is with the former. . . . The stars serve us scientifically only as providing positions.

---

5. *Light and Film* (New York: Time/Life Books, 1970), p. 50.

Elsewhere, Comte pointed out another "obvious" impossibility; we could never discover the temperatures of the heavenly bodies. He thus ruled out even the *theoretical* existence of a science of astrophysics; it is therefore doubly ironic that within half a century of his death, most of astronomy was astrophysics, and scarcely anyone was concerned with the solar system, which he claimed was "our only true interest."

The demolition of Monsieur Comte was produced by a single technological development: spectroscopy. I don't see how we can blame Comte for not imagining the spectroscope; who could possibly have dreamed that a glass prism would have revealed the chemical composition, temperature, magnetic characteristics, and much else, of the most distant stars? And can we be sure that, even now, we have discovered all the ways of extracting information from a beam of light?

Let us think about light for a moment, as the development of optics provides the most perfect example of the way in which technology can expand the frontiers of knowledge. Vision is our only long-range sense—unless one accepts ESP—and until this century everything that we knew about the universe was brought to us on waves of light. If you doubt this, just close your eyes, or reread H. G. Wells's most famous short story:

> Núñez found himself trying to explain the great world out of which he had fallen, and the sky and mountains and sight and such-like marvels, to these elders who sat in darkness in the Country of the Blind. And they would believe and understand nothing whatever he told them. . . . For fourteen generations these people had been blind and cut off from all the seeing world; the names of all things of sight had faded and changed . . . and they had ceased to concern themselves with anything beyond the rocky slopes above their circling wall. Blind men of genius had arisen among them and questioned the shreds of belief and tradition they had brought with them

from their seeing days, and had dismissed all these things as idle fancies, and replaced them with newer and saner explanations. Much of their imagination had shrivelled with their eyes.

With the genius of the poet he pretended he wasn't, Wells created in this story a universal myth: *"their imagination had shrivelled with their eyes."* And, on the contrary, how ours has expanded, not only with our eyes, but even more with the instruments we have applied to them!

Galileo and the telescope is the classic example; who has not envied him, for his first glimpses of the mountains of the moon, the satellites of Jupiter, the phases of Venus, the banked star-clouds of the Milky Way? During those few months in 1609–1610 there occurred the greatest expansion of man's mental horizons that has ever occurred in the whole history of science. The tiny, closed cosmos of the medieval world lay in ruins, its crystalline spheres shattered like the fragments of some discarded nursery toy. Which, in a sense, it was, being one of the childish things our species had to put aside before it could face reality. There will be other sacrifices to come.

One of the most remarkable things about the technology of handling light is the extreme simplicity of the means involved, compared with the far-reaching consequences. To give a humble but revolutionary example, consider spectacles.

The scholars of the ancient world, struggling to read by oil lamps and candles, must often have been functionally blind by middle age, especially as the manuscripts they studied were usually designed for art rather than legibility. The invention of eyeglasses (c. 1350) may well have doubled the intellectual capacity of the human race, for with their aid a man need no longer give up his work just when he is entering his most productive years. I don't know if this is an original idea, or whether it has already been refuted, but one could make a case for spectacles being a prime cause of the Renaissance.[6]

It is hard to think of a simpler piece of apparatus than a lens; yet what wonders it can reveal! Few people realize that the remarkable Dutch observer Van Leeuwenhoek discovered bacteria *with an instrument consisting of a single lens!* His "microscopes" were nothing more than beads of glass mounted in metal plates, yet they opened up a whole universe. Unfortunately, Van Leeuwenhoek was such a genius that no one else was ever able to match his skill, and the microscope remained little more than a toy. For its full use, it had to wait for Pasteur, two hundred years later.

And here is another of the great ifs of technology. Suppose Van Leeuwenhoek's observations had been followed up; then the germ theory of disease—often suggested but never proved—might have been established in the seventeenth century instead of the nineteenth. Hundreds of millions of lives would have been saved, and by this time the population explosion would have come and gone. Human civilization would by now have collapsed, or it would have safely passed through the crisis which still lies ahead of us—and for which, if you will excuse me, I have coined the word *Apopaclypse*.

The microscope and the telescope, both born about the same time, thus have sharply contrasting histories. The microscope remained a toy—the plaything of rich (well,

---

6. As one of the undoubtedly countless examples of the need for eyeglasses in the ancient world, see *Julius Caesar*, Act V, scene ii:

> Go, Pindarus, get higher on that hill;
> My sight was ever thick; regard Titinius,
> And tell me what thou notest about the field.

And if anyone wants to know what Romans would have looked like wearing horn-rimmed spectacles, Phil Silvers has already obliged in *A Funny Thing Happened on the Way to the Forum*.

fairly rich) amateurs like Samuel Pepys.[7] But the telescope, from the moment it was introduced, started a revolution in astronomy that has continued to this day.

Many years ago, however, the telescope came up against an apparently fundamental limit—that of *practical* magnifying power. Because of the wave nature of light, there is no point in using very high magnifications; the image simply breaks up, like an overenlarged newspaper block. Still, this natural limit is a very generous one. In theory, the Mount Palomar five-hundred-centimetre telescope would permit an incredible twenty thousand power, which would bring the moon to within fifteen kilometres.

Alas, this delightful fantasy is frustrated by the medium through which the light must pass—the few kilometres above the observatory. A star image can travel intact for a trillion kilometres, only to be hopelessly scrambled during the last microseconds of its journey by turbulence in the earth's atmosphere.[8] To the optical astronomer, all too often the medium is the mess.

Even under the rare conditions of virtually perfect seeing, at mountain-top observatories, the highest magnification that can ever be used is only about a thousand. This means that under favourable conditions the smallest object that can be seen on the moon is about half a kilometre across, and on Mars about fifty kilometres across. But these figures are very misleading, because contrast

---

7. On 26 July 1663 Pepys bought a microscope for the "great price" of five pounds ten shillings. "A most curious bauble it is," which he used "with great pleasure, but with great difficulty." Let us also never forget that, as president of the Royal Society, Pepys's name appears immediately below Newton's on the title page of the *Principia*.

8. Very recently, it has been possible to approach the theoretical limits of magnification by the technique known as "speckle interferometry." The five-hundred-centimeter Hale telescope has been used at an effective focal length of *four-fifths of a kilometre* to photograph the disks of giant stars like Antares.

plays a vital role. Lunar contrasts can be very high, owing to the starkness of the shadows. Mars contrasts are very low, making its surface features difficult to see and still harder to draw and photograph.

This tantalizing state of affairs led to one of the most famous, entertaining, and perhaps tragic episodes in the history of astronomy; I refer to the long controversy over the Martian canals, which has been finally settled only during the last few months. It is an example of what can happen when the desire for knowledge outruns the technology of the time.

Though he was not the first man to "observe" the canals, Percival Lowell was certainly the man who put them on the map—and I use that phrase with malice aforethought. Carl Sagan has, perhaps unkindly, referred to Lowell as "one of the worst draftsmen who ever sat down at the telescope"; I have preferred to call him "the man with the tessellated eyeballs."[9]

Whatever Lowell's deficiencies as an observer, there can be no doubt of his ability as a propagandist. In a series of persuasive books, from 1895 onwards, he almost singlehandedly laid the foundations of a myth which was gleefully elaborated upon by several generations of science-fiction writers, of whom the most celebrated were Wells, Burroughs, and Bradbury. The ancient seabeds, the vast irrigation system which still brought life to a dying planet, the ruins of cities that would make Troy seem a creation of yesterday—it was a beautiful dream while it lasted, which was until July 15, 1965.

On that day, the overstrained technology of the telescope was surpassed—though by no means superseded—by that of the TV-carrying space probe. Mariner 4 gave us our first glimpse of the real Mars, though by another delightful irony of fate those initial pictures were almost

---

9. Both libels will be found in *Mars and the Mind of Man* (New York: Harper & Row, 1973).

as misleading as Lowell's fantasies. Not until Mariner 9's superb mapping of the entire planet, in 1972, did Mars slowly begin to emerge from the mists as a unique geological entity—and one of our main orders of business in the next hundred years.

In the other direction, down towards the atom, we have also broken through one apparently insuperable barrier after another. The optical telescope and the optical microscope reached their limits at about the same time, since these are both set by the wave nature of light. The wholly unexpected invention of the electron microscope suddenly increased magnifying power a thousandfold, allowing us to view structures of molecular size and producing advances in the understanding of living matter that could have been obtained in no other way.

In the last few years, there has been another breakthrough. I hate having to use this exhausted word, but there are times when there is no alternative. The scanning electron microscope has done something quite new, and wholly beyond the power of the older optical and electronic instruments. By showing minute three-dimensional objects in sharp focus, it has allowed us for the first time to enter—emotionally, at least—the submicroscopic world. When you look at a good S.E.M. photo of some creature barely visible to the eye, you can easily believe that it is really as large as a dog—or even an elephant. There is no sense of scale; it is as if Alice in Wonderland's fantasy has come true and we are able to shrink ourselves down to insect size and have an eyeball-to-eyeball confrontation with a beetle.

The power of technology to change one's intellectual viewpoint is one of its greatest contributions not merely to knowledge but to something even more important: *understanding*. I cannot think of a better proof of this than some remarks made by Apollo 8's William Anders at the signing of the Intelsat agreement here in Washington on 20 August 1971:

Truly, the most amazing part of the flight was not the moon, but the view we had of the earth itself. We looked back from 240,000 miles to see a very small, round, beautiful, fragile-looking little ball floating in an immense black void of space. It was the only color in the universe—very fragile—very delicate indeed. Since this was Christmas time, it reminded me of a Christmas-tree ornament—colorful and fragile. Something that we needed to learn to handle with care.

Now, the telescope, the microscope, and even the rocket have given us only a change of scale or of viewpoint in *space*. In a more modest way, men have been achieving this ever since they started exploring the earth. What is very new in human history is the power to change our outlook on *time*.

The camera was the first breakthrough in this difficult area. From the beginning, the photographic plate could capture a moment out of time, in a manner never even conceived before—witness the quotation I have already given from that Leipzig newspaper. We all know the extraordinary emotional impact of old photographs; this is because they can provide, in a way not possible even to the greatest art, a window into the past. We can look into the eyes of Lincoln or Darwin, but not of Washington or Newton. At least, not yet.

Because the first photographic emulsions were very slow, the earliest glimpses of the past were somewhat extended ones; they lasted minutes at a time. But about a hundred years ago the camera acquired sufficient speed to provide mankind with another wonderful tool; call it an image-freezer, or time-slicer.

It is in fact almost exactly a century ago (1872) that the flamboyant Eadweard *(sic)* Muybridge solved a famous problem that had baffled every artist since the creators of the first cave paintings. Does a running horse have all four feet off the ground at the same time? Muybridge found that the answer was yes; he also discovered that the characteristic "rocking-horse" position shown in innumerable paintings of charging cavalry and Derby winners was nonsense. This caused great heartburning in artistic circles.

Since then, of course, the camera has speeded up many millionfold. Until recently, the ultimate was a photograph which Dr. Harold Edgerton has in his office at M.I.T. It shows a steel tower surmounted by a globular cloud with three cables leading into it. The ends of the cables are a little fuzzy, which is not surprising. The cloud is an A-bomb, a few microseconds after zero. . . .

Yet now we have far surpassed that. Using laser techniques, a slug of light less than a centimetre long has been stopped in its tracks. I suppose we'll reach the end of the line when someone catches a single photon in mid-vibration.

We are obviously a long way from the tempo of the running horse—just beyond the limits of human perception—or even of the hummingbird's wingbeat, which is still something that the mind can comprehend even if the eye cannot grasp it. Slicing time into thinner and thinner wafers has now led us into the weird world of nuclear phenomena, and perhaps even down to the atomic or granular structure to time itself. They may be "chronons," just as there are photons.

Now that every amateur photographer has a flash gun plugged into his camera, the power to freeze movement no longer seems such a miracle. And, of course, stopping time is not a *wholly* new experience to men, though in the past it was an uncontrollable one. A thunderstorm on a dark night was pre-twentieth-century man's equivalent of a modern strobe-light show, and must have impressed him even if he did not particularly enjoy it.

There may be worlds in existence which have natural strobe lighting, though until a few years ago not even the most irresponsible of science-fiction writers would have dared to imagine such a thing. For who could have dreamed of a star which switched itself on and off thirty times a second?

The Crab pulsar does just this, and it's strange to think that its flashes might have been discovered years ago, if anyone had been insane enough to look for them with suitable equipment. What would have happened to astronomy, I wonder, if that had been done? Many people would have been convinced that such a flickering star was artificial, and, even now, I don't think we should dismiss this explanation. The pulsars may yet turn out to be beacons, and those who protest that they are a very inefficient way of broadcasting have been neatly answered by Dr. Frank Drake. How do we know, he has asked, that there aren't some *stupid* supercivilizations around?

Granted the improbability that it could have survived the initial supernova explosion, it's fascinating to speculate about conditions on a world circling the Crab pulsar. To our eyes, daylight would appear to be continuous, but it would really be thirty cycles per second A.C. A rapidly moving object would break up into discrete images. Something that appeared to be stationary might be really spinning at high speed.

What would be the effect of this on evolution? Could predators take advantage of these weird conditions to deceive their victims? One day I may work this idea up into a story. Meanwhile, if Isaac Asimov (to take a name at random) uses it first, you'll know whom he stole it from.

So far I have talked about slowing down time, but what about speeding it up? Of course, that's much easier; it requires very simple technology, but lots of patience. Though the results are often fascinating, I do not know if they have yet contributed much to scientific knowledge. Time-lapse films of clouds and growing plants are the best-known examples in this field. Anyone who has watched the fight to the death between two vines, striking at each other like serpents, will have a new insight on the botanical kingdom. And I hope that the meteorologists will learn a great deal from the global cloud-movement films that have been taken from satellites; eventually these may give us a synoptic view of the seasons and even long-term climatic changes.

This was anticipated by two great writers of the last century:

> Night followed day like the flapping of a black wing . . . and I saw the sun hopping swiftly across the sky, leaping it every minute, and every minute marking a day. . . . The twinkling succession of darkness and light was excessively painful to the eye. Then, in the intermittent darkness, I saw the moon spinning swiftly through her quarters from new to full, and had a faint glimpse of the circling stars. Presently, as I went on, still gaining velocity, the palpitation of night and day merged into one continuous greyness . . . the jerking sun became a streak of fire, a brilliant arch, in space; the moon, a fainter, fluctuating band. . . . Presently I noted that the sun belt swayed up and down, from solstice to solstice, in a minute or less, and that consequently my pace was over a year a minute; and minute by minute the white snow flashed across the world, and vanished, and was followed by the bright, brief green of spring.

That, as I am sure you have all recognized, is from H. G. Wells's first—and greatest—novel, *The Time Machine*. Unlike his space romances, it is a book that could not have been written before the nineteenth century; only then had the geologists finally shattered the myth of Genesis 4004 B.C. and revealed the immense vistas of time that lay in the past—and may lie ahead. The emotional impact of that discovery on the more sensitive Victorians is preserved in Tennyson's famous lines in the poem *In Memoriam:*

> There rolls the deep where grew the tree.
> O earth, what changes hast thou seen!
> There where the long street roars hath been
> The stillness of the central sea.

> The hills are shadows, and they flow
> From form to form, and nothing stands;
> They melt like mist, the solid lands,
> Like clouds they shape themselves and go.

We now know that this poetic vision is a pretty good description of continental drift, suddenly respectable after languishing for years in the wilderness somewhere to the southeast of Velikovsky. And what established continental drift was a series of breakthroughs in technology, quite as exciting as those involved in the exploration of space. We are accustomed to sending probes to the planets. We have now begun to send probes into the past.

I don't mean this literally, of course; I don't believe in time travel, though I understand that Kurt Gödel has shown that it is theoretically possible under certain peculiar and highly impracticable circumstances, involving the annihilation of most of the universe. The best argument against time travel, as has been frequently pointed out, is the notable absence of time travellers. However, a few years ago, one science-fiction writer pointed out a chillingly logical answer to this. Time travellers, like radio waves, may need a receiver . . . and none has been built yet. As soon as one is invented, we may expect visitors from the future . . . and we had better watch out.

Looking into the past, however, does not involve logical paradoxes, and our time-probing has brought back knowledge which a few years ago would have been regarded as forever hidden. I wonder what Auguste Comte would have said if one had asked him the chances of finding the age of a random piece of bone, of locating the North Pole a million years ago, of measuring the temperature of the Jurassic ocean, or the length of the day soon after the birth of the moon? I feel quite certain that he would have said such things are as intrinsically unknowable as the composition of the stars.

Yet such knowledge is now ours, and often through

methods which, in principle at least, are surprisingly simple. Everyone is aware of the revolution in archaeology brought about by carbon 14 dating. The still more surprising science of palaeothermometry depends on similar principles. If you measure the isotope ratios in the skeletons of marine creatures, you can deduce the temperatures of the seas in which they lived. So we can now go back along the cores taken from the ocean bed, and watch the rise and fall of the thermometer as the ice ages come and go, one after another. There is, surely, something almost magical about this.

To track the wanderings of the earth's magnetic poles ages before compasses were invented or there were men to use them appears equally magical. Yet once again, the trick seems simple—as soon as it is explained. When molten rock cools, which happens continually during volcanic eruptions, it becomes slightly magnetized in the direction of the prevailing field. The tiny atomic compasses become frozen in line, carrying a message which sensitive magnetometers can decipher.

But what about the length of the day millions of years ago? Surprisingly, this requires the simplest technology of all; merely a microscope, and infinite patience.

Just as the growth of a tree is recorded in successive rings, so it is with certain corals. But some of them show not only *annual* layering but much finer bands of *daily* growth. By studying these, it has been discovered that six hundred million years ago the earth spun much more swiftly on its axis; it had a twenty-one-hour day, and there were 425 days in the year.

These remarkable achievements, and others like them, are allowing us to reconstruct the past like a gigantic jigsaw puzzle. Just how far can the process go? Is there *any* knowledge of the past which is forever beyond recovery?

A favourite science-fiction idea—though I have not seen it around recently—is the machine that can recapture images or sounds from the past. Many will consider that this is not science fiction, but fantasy. They may be right, but let us indulge in a little daydreaming.

There used to be a common superstition to the effect that no sound dies completely, and that a sufficiently sensitive amplifier could recapture any words ever spoken by any man who has ever lived. How nice to be able to hear the Gettysburg Address, Will Shakespeare at the Globe, the Sermon on the Mount, the last words of Socrates. But the naïve approach of brute-force amplification is, of course, nonsense; all that you would get is raw noise. Within a fraction of a second, all normal sounds, expanding away from their source at Mach One, become so dilute that their energy sinks below that of the randomly vibrating air molecules. Perhaps a thunderclap may survive for a minute, and the blast wave of Krakatoa for a few hours—but your words and mine last little longer than the breath that powers them. They are swiftly swallowed up in the chaos of thermal agitation which surrounds us; and when you amplify chaos, the result is merely more of the same commodity.

Nevertheless, there is a slight hope of recapturing sounds from the past—when they have been accidentally frozen by some natural or artificial process. This was pointed out a few years ago by Dr. Richard Woodbridge in a letter to the I.E.E.E. with the intriguing title "Acoustic Recordings from Antiquity."[10]

Dr. Woodbridge first explored the surface of a clay pot with a simple phonograph pickup and succeeded in detecting the sounds produced by a rather noisy potter's wheel. Then he played loud music to a canvas while it was being painted, and found that short snatches of melody could be identified after the paint had dried. The final step—achieved only after a "long and tedious search" —was to find a spoken word in an oil painting. To quote from Dr. Woodbridge's letter: "The word was 'blue' and

---

10. *Proceedings of the I.E.E.E.*, August 1969, pp. 1465–1466.

was located in a blue paint stroke—as if the artist was talking to himself or to the subject."

This pioneering achievement certainly opens up some fascinating vistas. It is said that Leonardo employed a small orchestra to alleviate Mona Lisa's boredom during the prolonged sittings. Well, we may be able to check this—if the authorities at the Louvre will allow someone to prowl over the canvas with a crystal pickup.

A few months ago I wrote to Dick Woodbridge to find out if there had been any further developments in this field, which, it should be pointed out, requires the very minimum of equipment; merely a pair of earphones, a phonograph pickup, a steady hand, and unlimited patience. But he had nothing new to report and ended his letter with a plaint which all pioneers will echo: "The bottom part of an S curve is a lonely place to be!" True, but there's a lot of room down there to maneuver.

There must be better ways of recapturing sound, but I can't imagine what they are. Still less can I imagine any way of performing a much more difficult feat—recapturing *images* from the past. I would not say that it is impossible in principle; every time we use a telescope, we are, of course, looking backwards in time. But the detailed reconstruction of ancient images—palaeoholography?—must depend upon technologies which have not yet been discovered, and it is probably futile to speculate about them at this stage of our ignorance.

It may well be hoped that—whatever the enormous benefits to the historian—such powers never become available. There is a peculiar horror in the idea that, from some point in the future, our descendants may have the ability to watch everything that we ever do.

But, because a thing is appalling, it does not follow that it is impossible, as the H-bomb has amply demonstrated. The nature of time is still a mystery; there may yet be ways of seeing the past. Is that any stranger than observing the center of the sun, which is what we are now doing with telescopes buried a mile underground? Surely neutrino astronomy, involving the detection of ghostly particles which can race at the speed of light through a million million kilometres of lead without inconvenience, is a greater affront to common sense than a simple idea like observing the past.

I seem to be in grave danger, just when I am running out of time, of starting on an altogether new talk: "Knowledge and the Limits of Technology." Which only proves, of course, how difficult it is to separate the two subjects—or to establish limits to either. In fact, no such limits may exist, this side of infinity and eternity. Those who fear that this is indeed true have often tried to call a halt to scientific research or industrial development; their voices have never been louder than they are at this moment.

Well, there may be limits to growth, in the sense of physical productivity, though in a properly organised world we would still be nowhere near them. But the expansion of knowledge—of information—is the one type of growth that uses no irreplaceable resources, squanders no energy. In fact, in terms of energy, information provides some almost unbelievable bargains. The National Academy of Sciences' recent report on astronomy gave a statistic that I would not credit from a less reputable source. All the energy collected by our giant radio-telescopes during the three decades that have revolutionised astronomy is "little more than that released by the impact of a few snowflakes on the ground."[11]

In the long run, the gathering and handling of knowledge is the only growth industry—as it should be. And to make the enjoyment of that knowledge possible, technology must play its other great role: lifting the burden of mindless toil, and permitting what Norbert Wiener called "the *Human* use of human beings."

We are only really alive when we are *aware,* when we

---

11. *Astronomy and Astrophysics for the 1970's* (Washington, D.C.: National Academy of Sciences, 1973), p. 77.

are interacting with the universe at the highest emotional or intellectual level. Scientists and artists do this; so, to the limits of their ability, did primitive hunters, whose lives we are now completely reassessing in the light of new knowledge. Anthropologists have just discovered, to their considerable astonishment, that we lost the twenty-hour week somewhere back in the Neolithic. For in optimum conditions, a few hours of hunting/foraging a day were all that was needed to secure the necessities of life; the rest of the time could be spent sleeping, conversing, chewing the fat (literally) and, of course, thinking.

But, as we have seen, thinking doesn't get you very far without technology. We can thank what I have christened the agricultural-industrial complex for that. Unfortunately, this sinister organization also invented work and abolished leisure, which we are only now rediscovering, after a rather nasty ten thousand years. It is to be hoped we will be able to make a safe transition into the postindustrial age, and then the slogan of all mankind will be— if I may change just two letters in Wilde's famous aphorism—WORK IS THE CURSE OF THE THINKING CLASSES.

It is technology, wisely used, that will give us time to think, and an unlimited supply of subjects to think about. And if it leads to our successors, either the intelligent computers or the "giant brains" that Olaf Stapledon described in his masterpiece, *Last and First Men*, why should that be regarded as a greater tragedy than the passing of the Neanderthals? Our technology, in the widest sense of the word, is what has made us human; those who attempt to deny this are denying their own humanity. This currently popular "treason of the intellectuals" is a disease of the affluent countries; the rest of the world cannot afford it.

Not long ago, I was driving through the outskirts of Bombay when I noticed a *sadhu* (holy man) with just two visible possessions. One was a skimpy loincloth; the other, slung round his neck on a strap, was a transistorized megaphone. There, I told myself, goes a man who does not hesitate to use technology to spread his particular brand of knowledge. He has grasped the one tool he needs, and discarded all else.

And that is the true wisdom—whether it comes from the East, or from the West.

23

To the Committee on Space Science

In July 1975 the House of Representatives Committee on Space Science and Applications, with Chairman Don Fuqua, held an extensive series of hearings on the future of the American space program. When it gave me a ticket to come all the way from Colombo to give evidence, I was both flattered and apprehensive. It was quite one thing to write inspirational prose about the wonders of space exploration in the centuries to come, but it would be quite another to answer even the friendliest interrogator wanting to know "Yes—but exactly *what* should we be doing in fiscal seventy-seven?"

My appearance before the committee took place on 24 July and (for my part at least) was a most enjoyable experience. To convey the flavor of the proceedings, I also give extracts from the subsequent question period.

The hearings were later published in three volumes, *Future Space Programs 1975,* which contain a wealth of information. The committee had cast its net widely, and the other nine witnesses who made personal appearances included *Saturday Review*'s Norman Cousins, Dr. Carl Sagan, NASA Administrator Dr. James Fletcher, and Dr. Gerald O'Neill, well known for his advocacy of space colonies. In addition there were over forty written contributions on widely varied subjects from such authorities as Isaac Asimov, Dr. Carleton Coon, Dr. Bruce Murray, Dr. Bernard Oliver, Dr. John Pierce, Dr. Edward Teller, and Dr. Wernher von Braun.

Some wit once defined an expert as "somebody from out

of town." So even in this distinguished company I could claim, at least on a mileage basis, to be the most expert expert summoned before the committee.

Mr. Chairman, distinguished members of the Committee:

I am deeply grateful for the honor of appearing here today. Despite my reluctance to leave my home in the beautiful island of Sri Lanka, your invitation was an offer I couldn't refuse, not only for the pleasure of meeting you all, but because of a rather strange literary coincidence.

I've just finished an ambitious novel, out early next year, at the climax of which my hero addresses the United States Congress and gives it some advice on future space programs. However, this event takes place at the Quincentennial, 301 years from now. . . .

May I therefore read you a few words from the *Congressional Record* for 4 July, 2276, which nevertheless seem relevant today:

> Members of Congress, let me first express my deep gratitude to the Centennial Committee, whose generosity made possible my visit to Earth, and to these United States. I bring greetings to all of you from Titan, largest of Saturn's many moons—and the most distant world yet occupied by mankind.
>
> Five hundred years ago this land was also a frontier—not only geographically but politically. Your ancestors, less than twenty generations in the past, created the first democratic constitution that really worked—and which still works today, on worlds that they could not have imagined in their wildest dreams.

Then follows a brief flashback to the four earlier Centennials. Herewith the one that most concerns us:

In 1976, the conquest of interplanetary space was about to begin. By that time, the first men had already reached the Moon, using techniques which today seem unbelievably primitive. Although all historians now agree that the Apollo project marked the United States' supreme achievement, and its greatest moment of triumph, it was inspired by political motives that seem ludicrous—indeed, incomprehensible—to our modern minds. And it is no reflection on those first engineers and astronauts that their brilliant pioneering effort was a technological deadend, and that serious space travel did not begin for several decades, with much more advanced vehicles and propulsion systems.

Then follows a proposal for a very advanced piece of space technology which I won't describe here, but which prompts some final rhetoric:

How those scientist-statesmen, Franklin and Jefferson, would have welcomed such a project! They would have grasped its scope, if not its technology— for they were interested in every branch of knowledge between heaven and earth.

The problems they faced, five hundred years ago, will never rise again. The age of conflict between nations is over. But we have other challenges, which may yet tax us to the utmost. Let us be thankful that the universe can always provide great goals beyond ourselves, and enterprises to which we can pledge our lives, our fortunes and our sacred honor.[1]

So much for the *Congressional Record* of 4 July 2276. I hope I have nicely disoriented your time sense, so that you are now free to take off into the future. . . .

Now, there are two aspects to the problem of technological forecasting. The first is the ability to see that some development is possible or desirable—preferably both.

---

1. *Imperial Earth* (New York: Harcourt Brace Jovanovich, 1976), chap. 41.

The second is to know when the time is ripe to do something about it, because it can be disastrous to be a premature pioneer. My own country provides some classic examples of this; the steamship *Great Eastern,* in the 1850's; the Comet jet airliner, a hundred years later. But not, I hope, Concorde.

From the nature of things it is very unlikely that one person can be qualified to give opinions in both these areas—that is, ultimate feasibility and immediate practicability. For myself I have always been more interested in the spectacular possibilities of the distant future, and not the practical problems of the day after tomorrow. Indeed, I've summed this up in the warning that if you take me *too* seriously, you'll go broke—but if you don't take me seriously *enough,* your children will go broke. And I'd like to add another warning which should be engraved on the desk of every committee chairperson in letters of gold—DON'T GIVE ALL YOUR BASKETS TO ONE EGGHEAD.

I've discussed the perils and problems of technical forecasting in *Profiles of the Future,*[2] where I've classified numerous past debacles under the headings "Failures of Nerve" and "Failures of Imagination." The "Failure of Imagination" prompted "experts," only a lifetime ago, to pronounce that heavier-than-air flight was impossible, and space travel was not even worth discussing.

This particular failure is less common nowadays, because we have seen so many wonderful achievements that the public is prepared to accept almost any miracle of science or technology. In fact, the pendulum has swung too far the opposite way—towards overcredulity. Hence the unfortunate popularity of fraudulent or downright insane books about aerial crockery, antediluvian astronauts, emotional cabbages, Bermuda hexagons, and so forth, ad nauseam.

Perhaps we are more afflicted now by the failure of

---

2. New York: Harper & Row, 1973.

nerve: the appreciation that something is certainly possible, coupled with the assertion that it is too far ahead
to be of any practical concern. This question of timing is
the most difficult one in the whole area of forecasting, because it can result in total disagreement among authorities. Perhaps the best example of this occurred in a
committee room not a stone's throw from here. Listen to
Dr. Vannevar Bush in 1945:

> There has been a great deal said about a 3,000
> mile . . . rocket. In my opinion such a thing is im-
> possible for many years. The people who have been
> writing these things that annoy me have been talking
> about . . . a rocket . . . so directed as to be a precise
> weapon which could land exactly on a certain target,
> such as a city. . . . I feel confident that it will not
> be done for a very long period of time to come. . . .
> I think we can leave that out of our thinking. I wish
> the American public would leave that out of their
> thinking.[3]

Well, the ICBM was then only ten years in the future.
Yet there are few scientists to whom this country owes
more than the late Dr. Bush, and in the total record I
would not care to match my crystal ball against his.

It's a cliché that we often tend to overestimate what
we can do in the near future—and grossly underestimate
what can be done in the more distant future. The reason
for this is very obvious, though it can only be explained
with a certain amount of hand-waving. The human
imagination extrapolates in a straight line, but in the real
world, as the Club de Rome and similar organisations are
always telling us, events follow a compound-interest or
exponential law. At the beginning, therefore, the straight
line of the human imagination surpasses the exponential
curve; but sooner or later the steeply rising curve will
cross the straight line, and thereafter reality outstrips
imagination.

---

3. U.S. Senate, *Hearings,* December 3, 1945.

How far ahead that point is depends not only on the difficulty of the achievement, but also upon the social factors involved. Let me give an example from my own experience.

It is exactly thirty years ago this month that I wrote my paper on communications satellites.[4] One of the reasons that I never attempted to patent the idea, apart from sheer laziness, was I simply did not expect comsats to be realised in my lifetime. . . . Yet Early Bird was only twenty years ahead.

Why was this particular development so extraordinarily swift? Because a primary human need—that of communications—was involved. In what was, historically, a mere blink of an eyelid, the TV set left the lab and invaded every home, even the poorest. When the human race sees something it must have—and news, entertainment, information of all kinds come high on that list—it insists on having it, whatever the cost. All over this planet, TV antennas rear above squalid shacks. And, conversely, men show much less enthusiasm for other things that may be logically just as desirable, but do not have the same emotional appeal.

Thus it may well be argued that the Earth Resources Satellite, and the meteorological satellite, have a potential economic impact as great as that of the communications satellite. But because they lack its glamour—they won't bring Muhammad Ali live from Kuala Lumpur—it is not so easy to convince the skeptical taxpayer of their value.

So much for generalities. Now I would like to come down to specifics, selecting just a few items from the enormous range of space possibilities.

I am rushing straight back from this hearing to Asia because, as you are aware, one of the greatest educational experiments in history is due to start on 1 August—the use of the ATS-6 direct broadcast satellite to distribute TV to several thousand receivers scattered over India.

---

4. "Extra-Terrestrial Relays," *Wireless World,* October 1945.

Although the primary purpose of the experiment is to get information on family planning, hygiene, and agricultural techniques into remote villages, it is also hoped that no less than a quarter of all the teachers in India will be exposed to special training programs that will upgrade their efficiency. Thus even this limited one-year experiment could have far-reaching consequences, and it will be watched with great attention by the whole world.[5]

I'm delighted to report that the Indian Space Research Organisation is very generously shipping me an entire ground station so that I can set it up in my Colombo house, to demonstrate the programs to all the local educators and communications engineers. During the next few months, therefore, I hope to show my friends in Sri Lanka, which as yet has no TV service, the possibilities of this new medium.

I would like to see educational satellites spread over the world as swiftly as the communications satellites have done, but I realise that the problems are much greater. Whether or not in this case the medium is the message, there is no doubt that success or failure will depend entirely on the software: the program content. That is expensive, but the potential audience is so enormous— *billions* rather than millions, over the course of years— that the investment would soon pay off handsomely. There may be no other way in which whole nations can be brought into the modern world within a single generation.

H. G. Wells once remarked that future history would be a race between education and catastrophe. "Edsats" could help us to win that race. Can you imagine Sesame Streets for the whole world? You may smile, but in many areas of knowledge it could be done. Despite the obvious problems of language, there are some subjects—mathematics, basic science—where a vast amount of informa-

---

5. See "Schoolmaster Satellite," read into the *Congressional Record* for 27 January 1972 by Representative William Anderson. (Chapter 12.)

tion can be put across by sound and vision alone, without the use of words.

Some experts believe that the cost of providing direct broadcast TV education to a medium-sized country would work out at around a dollar per pupil per year. That's only the cost of the satellite and the TV sets; the software and personnel would work out at several times as much. Nevertheless, one can envisage an educational satellite system covering the whole world—the educational equivalent of Intelsat—running for a few billion dollars a year.

It's hard to think of a more challenging or inspirational prospect. Just suppose that the United States—one of the two countries with the capability of providing the hardware—offered it to the entire developing world, and not merely to India for a single year? I happen to know that, before its recent convulsions, the White House was considering some communications spectacular in connection with the Bicentennial. Even now that might not be too late.

In any event, direct TV broadcasting from space is inevitable, especially for countries that have no alternative, with all that this implies in terms of free flow of information across frontiers, and the abolition of today's artificial barriers. (As the debates on the subject at the United Nations have shown, some people are very worried about this.) Before the end of this century, the communications satellites will have decided whether English, Russian, or Chinese is the second language of mankind.

In August 1971 I was asked to address the ambassadors of the nations gathered at the State Department to sign the agreement setting up Intelsat. I'd like to quote some comments I made then.[6]

I believe that communications satellites can unite mankind. Let me remind you that, whatever the his-

---

6. U.S. Department of State brochure, August 20, 1971. For complete text, see Chapter 12.

tory books say, this great country was created little more than a hundred years ago by two inventions. Without them, the United States was impossible; with them, it was inevitable. Those inventions, of course, are the railroad and the electric telegraph.

Today we are seeing, on a global scale, an almost exact parallel to that situation. What the railroads and the telegraph did here a century ago, the jets and the communications satellites are doing now to all the world.

I hope you will remember this analogy in the years ahead. For today, whether you intend it or not— whether you *wish* it or not—you have signed far more than just another intergovernmental agreement.

You have just signed a first draft of the Articles of Federation of the United States of Earth.

I don't believe this is mere hyperbole. What we are seeing now—largely as a result of space technology—is the establishment of supernational, global-service organisations in which all governments, in their own sheer self-interest, will simply *have* to cooperate. Intelsat is the obvious prototype. The World Weather Watch, in which meteorological satellites are an essential element, is another. And the dramatically successful Earth Resources Satellites, Landsat 1 and 2, may be the precursors of a sort of global inventory, not only surveying all our planet's resources, but also monitoring their misuse through pollution and environmental degradation. Our children will be unable to understand how we ran our world without these tools of space. The answer, of course, is that we didn't.

I'll say no more about application satellites, because there can now be little argument about their value, only about the priority given to each type. Let's move on now to the more controversial subject of scientific space exploration and manned space flight.

It is almost impossible to overpraise the achievements during the last decade of the space probes, Ranger,

Orbiter, Mariner, and Pioneer, which have revolutionised our knowledge of neighboring worlds. Let us hope that the forthcoming Viking missions—potentially, the most exciting of all—are equally successful. But how many people know about the discoveries that these robot explorers—our scouts into the new wilderness—have radioed back from Mars, Mercury, Venus, Jupiter? It's a great tragedy that the United States has lost the most effective medium for spreading news of these adventures in science—the glossy magazines like *Life* and *Colliers*, which played such an important role in promoting the Space Age. The pictorial riches stored up in NASA are almost unknown to the general public; apart from a few specialised periodicals, the *National Geographic* is now about the only display forum for this superb material. And though the *Geographic* does a magnificent job for a large and influential audience, it hasn't the vast coverage of the old *Life*. I don't know what, if anything, can be done about this public relations problem, which concerns the whole of science, and not merely astronautics. If it could be solved, the deep-space research program would be in good shape.

With regard to manned space flight, we have at least settled one old controversy. I can remember the time, not long ago, when it was solemnly asserted that men could not survive zero gravity for more than a few hours. Well, we've now seen that they can function happily in space for months; indeed, it appears that some would be happy to stay there indefinitely, as long as their wives or girl friends could join them.[7]

During the last year there has been much publicity for the idea of "colonies in space," and it has even been suggested that they could provide a solution to some of today's problems. This idea is very old. It was developed in remarkable detail by the great Russian pioneer

---

7. At this point a protest from the audience forced me to add "husbands and boy friends."

Tsiolkovsky as early as 1911[8] and by J. D. Bernal in the 1920's.[9] I have not the slightest doubt that such schemes will eventually be realised. Indeed, many years ago I stated that *in the long run* there will be more people living off the earth than ever lived on it.

But that is looking centuries ahead, and I don't believe that we should concern ourselves with vast space cities this side of 2001. The technical problems are so enormous —there are so many possibilities for disaster owing to some trivial oversight or violation of ecological principles—that we must first prove we can make cities work down here, before we design Astropolis. I'm almost tempted to quote Vannevar Bush again: "I feel confident that it will not be done for a very long period of time to come. . . . I think we can leave that out of our thinking." I may be as wrong as Dr. Bush.

What we *should* start thinking about now are space villages, not space cities. We will need them in the quite near future, for the industries and services that will undoubtedly be established in Earth orbit. Incidentally, it's amusing to note that the very first suggestion for a manned satellite was made a little more than a hundred years ago by none other than the chaplain of the Senate, Edward Everett Hale, in his story "The Brick Moon."[10]

Even today, some of the reasons why we'll need men in orbit are quite obvious. Much complex scientific equipment can only be assembled, checked, and refurbished if there are men on the spot to supervise operations. The large space telescope is an excellent example. A permanent manned crew isn't necessary—indeed, in this case it's actually undesirable—but access from time to time is

---

8. "The Investigation of Universal Space by means of Reactive Devices." NASA Technical Translation TT F-243.

9. *The World, the Flesh and the Devil,* 1929 (Bloomington: Indiana University Press, reissued 1969).

10. *His Level Best and Other Stories* (Boston: Roberts, 1872).

essential. Much of the cost of today's satellites stems from the fact that they have to be built for absolute reliability, and have to perform complicated deployment maneuvers automatically after they are launched. It's been said, with perhaps only slight exaggeration, that there are billions of dollars of useless satellites in orbit right now which might be fixed by men with screwdrivers. To quote Churchill, this is nonsense up with which we shall not put.

One dramatic possibility, which doubtless has already been presented to your committee, is that of the orbiting solar power plant, several kilometres on a side, which would obviously have to be assembled over a long period of time by manned crews. However, it is very hard to believe that this could compete economically with ground-based installations in desert areas; but one day we will certainly have to go into space to tap the limitless energies of the sun. When that day comes we may go much closer, say to Mercury, where there is ten times more power available per square metre than in Earth orbit. Moreover, Mercury, which is much denser than the moon, probably contains vast quantities of metals. So *that* is where we may locate the heavy, polluting industries of the centuries to come, shipping the finished products back to Earth. If this sounds fantastic, please remember that the cost of transporting material *across* the solar system is quite small even with present techniques; it's getting *away* from Earth that's expensive. This won't always be the case; I'll return to that theme later.

As our civilisation becomes more and more reliant upon applications satellites of ever-increasing power and complexity, it will eventually be essential to have permanent servicing and maintenance teams in orbit probably in the thirty-six-thousand-kilometre-high synchronous orbit, where most of the action will be). This state of affairs will arise before the end of the century as a result of *foreseeable* developments. It may be accelerated by discoveries in zero-gravity processing, space medicine, and so forth.

And apropos of space medicine, I would like to mention a cautionary fable I wrote exactly fifteen years ago. By another odd coincidence it takes place just around now, much of it in this very building. It concerns a United States senator who, at a 1975 hearing of the Committee on Astronautics, shot down the NASA administrator so effectively that he couldn't get the funds to build a manned space station. As a result, the Russians pioneered space hospitals where, among other things, heart patients could recover under the stress-free conditions of zero gravity. You can doubtless guess the sequel: the senator found *he* had a heart defect, and though the Russians were willing—indeed, delighted—to treat him, he realised that he had morally forfeited the right to avail himself of the still limited facilities. Hence the story's title, "Death and the Senator."[11]

Now, I said that this piece of space propaganda was a cautionary fable, and even in 1960 I didn't intend it to be taken as serious prophecy. But I *am* serious about the underlying message; there will be countless direct human benefits from space research which we cannot anticipate today; and if we do not keep our options open, we shall miss them or, what is almost as bad, be unable to exploit them when the time is ripe.

Hence, of course, the importance of the shuttle, of which this committee has heard a great deal, and will be hearing a great deal more. It's unfortunate that the

---

11. *Tales of Ten Worlds* (New York: Harcourt Brace Jovanovich, 1962). The same volume contains another cautionary tale, "I Remember Babylon," about an attempt to brainwash the United States by pornographic TV programs from a direct-broadcast satellite. This story, originally published in *Playboy* in May 1960, also quotes from my written evidence in the report, *The Next Ten Years in Space,* to the House of Representatives Committee on Astronautics in 1959.

shuttle, once touted as the DC-3 of space, has now been degraded for fiscal and other reasons to the DC-1½. But it's the only shuttle we have, and perhaps the only one anybody is likely to have in the eighties, so we must make the most of it. If all goes well, it should provide the final convincing demonstration of the need for men in space, not just on occasional sorties but as full-time workers. And this in turn will create a demand for cheaper and better methods of space transportation—exactly as happened with aviation.

Even according to the most optimistic estimates, the shuttle, though an essential intermediate stage, is still many orders of magnitude too expensive. Yet in terms of mere energy, space travel could be one of the cheapest forms of transport since the old sailing ships. The basic cost, in kilowatt hours, of lifting a man completely away from Earth is ten dollars; the cost of shipping him around the solar system is about the same. In principle, the communications and life-support equipment should cost far more than the propulsion. . . .

There is no way of getting anywhere near these theoretical figures with today's or tomorrow's technologies. We need new scientific breakthroughs; we've been using up our store of fundamental knowledge at a dangerous rate, and are running out of know-how. In fact, there have been almost no really new ideas in the field of astronautics for the last fifty years; you'll find practically all the things we are doing today outlined, at least in principle, in the classic works of Tsiolkovsky, Goddard, and Oberth, who, incidentally, celebrated his eighty-first birthday last month.

How will we find the new ideas we need? In the usual way: by keeping our eyes open, and by programmed luck, or serendipity. We must always have some people working on far-out, even apparently crazy, concepts. And working by themselves, in their own time, and without having to produce progress reports every six months. For as someone once remarked, if there had been government departments of scientific research in the Stone Age, by now we'd have had absolutely marvellous flint axes and arrowheads, but nobody would have invented steel.

I'll just mention two advanced ideas that may intrigue you. If I had a dollar to invest in them, I'd put ninety-five cents on the first, and three on the second, keeping the other two until something even crazier came along.

Fifteen years ago, two of the world's greatest theoretical physicists, with the formidable backing of Werhner von Braun, Theodore von Karman, General Curtis LeMay, Harold Urey, Arthur Kantrowitz, and others, spent some $10 million on Project Orion. This was a scheme to launch really large payloads—hundreds of tons—by the use of small nuclear explosions reacting against a pusher plate and a shock-absorbing system. Calculations and flight tests using chemical explosions proved that the project, incredible though it sounds, would really work, and that it would be possible to send large, lavishly equipped expeditions to any of the planets, by the expenditure of a thousand or so low-powered fission bombs. Quite a small fraction, say, of NATO's stockpile.[12]

When the project was killed by the Test Ban Treaty of 1963, Professor Freeman Dyson wrote an indignant article stating that "this was the first time in modern history that a major expansion of human technology had been suppressed for political reasons."[13]

Well, it seems to me that the reasons were rather more than political, for I find it hard to get enthusiastic about any vehicle which leaves a trail of a thousand exploding A-bombs behind it. But now there is at least the theoretical possibility of a *clean* Orion project, virtually free of radioactivity. The work that is going on at the moment to trigger fusion reactions with laser pulses would seem ideally suited to such a propulsion system. One can

---

12. John McPhee, *The Curve of Binding Energy* (New York: Farrar, Straus & Giroux, 1974).

13. *Science*, July 9, 1965.

imagine microspheres of hydrogen-deuterium being zapped several times a second during the climb through the atmosphere, and at more leisurely intervals thereafter.

Something of this kind may be demonstrated in the laboratory during the next few years; I am sure that there are some very bright people working on it right now, but I've seen nothing in print. (Maybe *I* shall be zapped as soon as I leave this hearing.) If feasibility can be proved, then a truly exhilarating prospect will open up before us, perhaps at the turn of the century. Space travel will no longer be propulsion limited; we can contemplate the large-scale exploration of all our planetary neighbors.

Now for the crazy idea.

In 1966 a group of oceanographers led by Dr. John Isaacs of Scripps described a true "skyhook."[14] They pointed out that it would be theoretically possible to lay a cable from a satellite in geostationary orbit *all the way down to the surface of the earth!* And then, in principle, one could send payloads up the cable by simple mechanical means. An electric elevator to space, or a Streetcar Named Heaven. . . .

I must confess that I didn't regard this as more than a scientific pipe dream until a couple of years later, when Cosmonaut Alexei Leonov gave me a copy of his handsome book, *The Stars Are Waiting*, published in 1967. Imagine my surprise when I saw that the Russians had come up with the same idea quite independently—the space elevator! (Poised, I've just noticed, immediately above Sri Lanka, though presumably the cable ends in Africa, since an equatorial site is mandatory and we're seven degrees north.)

What's the snag? Well, we need far stronger materials than those known today to make the scheme practicable. And there are many other problems; if anything went

---

14. *Science*, February 11, 1966.

wrong, we'd have a thirty-thousand-kilometre-high structure crashing down on our heads. But it's certainly worth thinking about, and it would be just the thing to carry down the electricity from that orbiting solar power plant.

But the *real* breakthrough which will open the gate to the universe probably lies in some direction no one can imagine today, any more than, a hundred years ago, one could have conceived of atomic energy. There's a feeling in the air now that something is happening to the foundations of physics; the animals from the nuclear zoo are escaping in all directions. New particles are discovered every month; perhaps there are new forces also awaiting discovery, and from them may come at last the old dream of controlling gravity. I'd put my remaining two cents on *that*.

Finally, a glance at remoter horizons. Less than a year from now, the first American Mars lander, Viking 1, will descend through the thin atmosphere of our neighbour—accompanied or preceded, no doubt, by several Marskods.[15] The Soviet Mars program has been much more ambitious than the United States one, and much less successful. But it has earned success, and I hope achieves it. There is plenty for everyone to discover on a world with almost the same land area as Earth.

There now seems little doubt that suitable life forms could exist, and even flourish, on Mars. We will be very, very lucky if Viking—or its Russian comrades—detect life twelve months from now. But that possibility exists; if I hadn't spent my dollar, I'd put, oh, five cents on this happening in 1976. And I'd put fifty on life being discovered on Mars *eventually*. Not intelligent life; though even that cannot be ruled out.

The impact of such a discovery could be as great as that of the first Sputnik. It will certainly change the

---

15. The Russians let me down. This was the first occasion they failed to launch Mars probes. (But they were doing very nicely on Venus.)

priorities of space exploration. At the very least, it will mean that the whole scientific establishment will jump on the manned space-flight bandwagon, instead only a section of it—though an ever-growing section—since the brilliant Skylab missions.

And what about *intelligent* life? In view of man's remarkable inability to get along with himself, perhaps it's just as well that there are probably no other rational beings in this solar system, at this moment of time. But that they must exist somewhere is now doubted by very few scientists. The only disagreement concerns the possible methods of detecting them. They will already have detected *us;* our radio emissions now fill a sphere containing hundreds of stars.

Doubtless, many of you have seen the NASA-Ames Project Cyclops report, outlining the technology for a very large radio-telescope and the associated signal-processing equipment, which would give a good probability of detecting intelligent signals from space.[16] Cyclops would inevitably contribute so much to the development of radio astronomy that it merits building for its own sake, and it could be an internationally funded, global project that would challenge the imagination of all mankind. There may be no other way in which we can discover —or perhaps establish—our position as intelligent beings in the hierarchy of the universe.

For this, in the long run, is what space exploration is all about. And that is why those of little courage, or little imagination, are so often opposed to it. A few weeks ago, I am sorry to say, a British minister who shall be nameless (actually he has several names) held up the first sample of North Sea oil and declaimed that it was more important than the moon shot, which "only brought back soil and rock." Well, even for a British minister, whose attention span seldom extends beyond the next Cabinet reshuffle, this was a singularly short-sighted remark.

---

16. NASA–Ames report CR 11445.

Everyone knows that the North Sea oil will be gone in a generation. But the moon is the first stepping stone to the riches of the whole universe.

The greatest lesson that we can draw from space is one of hope. In the absolute sense, as far as we can see into the future, there are *no* real limits to growth. It is true that we must cherish and conserve the treasures of this fragile earth, which we have so shamefully wasted. But if we come to our senses in time, we may yet have a splendid and inspiring role to play, on a stage wider and more marvelous than ever dreamed of by any poet or dramatist of the past. For it may be that the old astrologers had the truth exactly reversed, when they believed that the stars controlled the destinies of men.

The time may come when men control the destinies of stars.[17]

SUBSEQUENT QUESTION PERIOD

Mr. Fuqua: Mr. Clarke, what are some of the criteria that you think that we, as members of Congress, in trying to respond to proposals for future space programs . . . might use to try to analyze these programs?

Mr. Clarke: Of course, this is the fundamental problem, and all one can say on this is that you should get advice from the most informed experts you can, use common sense to analyze it, but don't let yourself be blinded by science. Remember what President Kennedy said after the Bay of Pigs: "The advice was unanimous and the advice was wrong."

Also, I think you should try to concentrate on things which are now of immediate applicability, which can be demonstrated even to the man on the street who doesn't

17. The epilogue to *First on the Moon* (Boston: Little, Brown, 1970).

know anything about science, or may even dislike science, as I'm afraid many people do.

Fuqua: This was my next question: How do we try to arrive at that delicate balance between applications to down-to-earth problems, as you referred to them, that the man in the street understands or most readily understands, and those that—and rightfully so—advocate that we should be doing more in the area of space sciences?

Clarke: These are always educational problems ultimately. You've got to really help people understand what you're saying, and you can best do this by giving specific examples which will appeal to everybody. I'm hoping if this Indian experiment succeeds—and everything depends on the software and the politics—this may with luck present to the world a way in which a nation with much worse problems than this country can use space technology to solve some of those social problems. If this works out, this would be of great value to the space program.

As far as trying to convince people of the value of pure scientific research—well, I mentioned that we're running out of know-how, and this is true, and much of the know-how which you get from scientific research isn't valuable for decades.

The example I like to give to people who ask me this question is this: In the 1890's a scientist was working on something of no conceivable importance, the conduction of electricity through gases, and I'm sure his friends must have said to him why don't you do something useful, like try to find the cure for the common cold? And he found X-rays, perhaps the greatest medical discovery of all time. The whole technology of medicine was transformed.

But I think people could understand the value of pure science if you can put across a few specific examples that hit them personally.

Mr. Winn: I was very interested in the fact that you discussed public relations problems, which concern the whole of science. This is something that the committee has been deeply concerned about for some time, on how we do a better job in America of selling the taxpayer on our requests for funds.

I just wondered if you could give the committee any of your ideas, being closely affiliated with the field, on how we could improve what we're doing at the present time. And I'm sure you would do it without being critical of any of the organisations, but if you want to be critical, be critical. That's all right with us.

Clarke: Well, Mr. Winn, it so happens that I have been for the last five years trying to do just this. I spent about $400,000 doing a TV documentary called "The Promise of Space," which we tried to show on the air here. We went to India and filmed this. We showed all the applications of space technology. Everyone who has seen this has said it's marvelous. Some of you may have seen it, or part of it, but we've never been able to get it on the air. There has been a tremendous resistance in the last few years even to listening to the facts on space technology.

Winn: Why?

Clarke: People just don't want to hear it, and why this is I don't know. Maybe it's part of the general climate of opinion and the difficult times this country's been going through, Vietnam and Watergate and all these things. I hope we're pulling out of this now. In fact, this interest in Professor O'Neill's program, which seems to be very great—although I don't take it very seriously in the short-term view—I think is one of the few encouraging things I've seen in the space field recently.

Winn: But do you have any other ideas? We have had all kinds of television programs, short inserts that NASA has tried, encouraging the media to use in the news. . . . I don't know that people would listen to an hour show on space, but they might.

Clarke: Well, there was a two-hour show on public broadcasting the other night. That did essentially, I think, not as good a job as we did in one hour, but at least they were able to get theirs on the air.

Winn: Well, I want you to know that the committee is definitely concerned about it because we don't seem to be doing the job. We have talked to NASA about it and they have changed some of their directions but we still don't seem to have been able to come up with the idea. You talked about Professor O'Neill's space-colonization concept a little bit. What is your own personal opinion of that?

Clarke: I'm sure that that sort of thing will be done eventually; even more ambitious things will be done eventually. But when we cannot design a piece of relatively conventional engineering, like the Concorde or an SST, without 1,000 percent cost overrun, and all sorts of unforeseen problems, the idea of building whole communities, whole worlds from scratch, at this stage I think is premature. Certainly, design studies, theoretical studies, should continue, but I don't think that anyone will be doing anything on this scale until well into the next century.

Winn: Of course, he scared some of us with the amounts of money. I don't know if you saw copies of his testimony, but they ran up as high as $200 billion.

Clarke: My gut reaction is that he's still off by several zeros.

Winn: You figure there'll be a cost overrun for space colonization?

Clarke: Yes.

Mr. Lloyd: Mr. Clarke, first of all I, too, would like to say how very pleased we are that you took your time and energy and efforts to join with us in educating some of us, who hopefully are educatable but haven't very much background in the development of Space Age technology and the problems of getting there and the raising of money, and all the rest of it. Do you feel that the United States at the present time is within the realm of reason, relative to the total gross national product with the money we have available? With the economies which are presented to us at the present moment, are we indeed giving

the proper amount of attention towards the things that
you are talking about?

Clarke: Well, frankly, I'm not familiar with the exact
figures since, as I say, it's two years since I've been in
the States. It would be interesting to compare them with
other countries. The United Kingdom's, I believe, are too
low—or so my friends say. The U.S.S.R.'s I think are
several times that—and I know it's very hard to get those
figures—but my feeling from what I have seen is that
the percentage is still too low. You have this problem of
running out of know-how, running out of theoretical
bases on which our future development does depend.

Lloyd: Well, let me address myself to that just for a
moment. As you know, our military budget this year is
roughly $100 billion. Obviously, any effort in this specific
area that we are addressing isn't going to be remotely a
portion of that.

Just addressing yourself to this, how could we possibly
take some of the money that we pour into our military
budget, $16 billion for a Trident, and so on and so forth . . .
how can we, from a public relations or a public point of
view, take some of this very defensive attitude and shift
it over into a more broad, open attitude of space-
mindedness?

Clarke: The Apollo-Soyuz experiment, of course, is
hopefully the beginning of some such relaxation. Getting
money out of the military budget into this—there is one
way of doing it. It's a tricky one, of course, but the military depend on research almost more than anybody else,
because they know well, although they might not like it
sometimes, that research may wipe out some whole
weapons systems. You've got to have something going on.
The laser is a good example of this. I mean, we suspect
that the laser is going to transform war, and also, as I
said earlier, lasers may transform space propulsion. I
guess we had better make sure that the military have the
money to do the research on lasers that they need.

I am, I guess, a kind of pacifist myself, even though I

was in the Royal Air Force, and I'm unhappy about things being done this way. But this, of course, is how space travel began, and how the rocket was funded originally, however deplorable it may be. So, I guess you must make sure that the military has the money for research, and then you must try to make sure that the results of that research are declassified as quickly as possible.

Mr. Frey: Another interesting thing, I think, is the paradox that we face, that the technology that has done so much for us has become the bad guy today in our society. We blame a great deal on it. Of course, technology is a neuter; it really isn't good or bad, it's how we use it. And science is somewhat the same way. In our engineering schools in this country we have had a tremendous drop-off of people going into the field, compared with what happened after the Sputnik. It seems that you're really looking at and talking about the mores of a society and the climate of opinion.

It's a frustrating thing, because there's so much payoff in space. It is as if we have been working for so long to build the base of the pyramid, and now we're about to build on that foundation and really come up with so much. Yet we're fighting a situation where no matter what the Administration, we get less money. There is less enthusiasm in the Congress, and there is less enthusiasm throughout the country.

It's my feeling—and I hope you'll disagree with me—that we are really not going to be able to change this until we have a tremendous problem that comes down and hits us on the head, or until people can look up and see the shuttle and see directly what it does. It looks to me like we're going to have even more of a problem in the next five years, trying to push for the things we believe in, than we did before. I'd just like your comments on these thoughts.

Clarke: It does seem that many of these problems—we all agree on the problems—look insoluble. I keep on mentioning education. I'd like to mention one other thing.

How many engineers and scientists do you have in Congress at the moment? I believe about one-third of the Presidium of the U.S.S.R.—I'm not sure of that—are engineers and scientists. I hope you'll excuse me speaking as a non-American, but one thing you've got to try to do is to get more engineers and scientists in Congress, and for obvious reasons. Until you do this, you have an unbiased representation of society. I think that you gentlemen do a remarkable job of coping with scientific and technical problems, but inevitably someone who's actually in the area of scientific expertise can see the problems more easily, just as it's very difficult for a scientist to understand legal problems and all the ramifications that may be involved there. So, this is one recipe for improving the situation, I think.

Frey: But what do you do? How do you excite the imagination of the people about, as you point out, the earth's resources, these things that are benefiting life today and will do so even more?

Clarke: Well, if you'll excuse the commercial, one of the ways of interesting people, young people, is science fiction. I've been saying this for years. I know an awful lot of scientists, and astronauts for that matter, who were turned on by science fiction. This is part of the educational process.

One reason is that science fiction is usually optimistic. Much modern fiction is very pessimistic, and in the science-fiction area you usually do have the feeling that the future can be better than the past, we can do something about the future, we don't have to wait until it comes and clobbers us.

Frey: We have many young Americans who don't see any limits on their horizons. I guess that's really what's going to get us through this period of time more than anything else.

Mr. Downing: I agree with you we ought to have more

scientists and occasionally we do. But our problem is we can't get a scientist who is also a politician. We had one on this committee and he didn't last but one term. . .

I have a vague question following up Congressman Frey. I want to know how, in your opinion, do we best sell the space program to the American taxpayer, as we go into the less-dramatic phases of the space program? It's easy to see why he would react favorably after he saw Sputnik. Thank goodness he saw Sputnik and became concerned about it, the military implications, and the dramatics of putting a man on the moon in ten years. But now we're on a plateau, where it's pure science, research, and here interest in the program is bound to wane. The problem is how do we sell it in order to keep the space program going?

Clarke: One thing will help you: when the first real space products start hitting the market. There are some already, of course, but they're rather esoteric—computers and calculators, some aspects of space medicine, heart pacemakers—there are many of them. The space program has probably already paid for itself. But you have to do a lot of cost accounting to prove that. It's not an emotional thing.

But when people start seeing—as I think they will in the next decade—products which are literally made in space, then they'll begin to understand it. It's going to be an uphill job. There's no quick answer. I can't give you a formula now that will solve the problem. As you say, we're on a plateau, and that might extend maybe another five years into the future.

Downing: Well, I think people are accepting technologies from the space program, without realizing they come from the program. Several years ago I had the privilege of being on the *Queen Elizabeth* and the commodore invited us to the bridge. One lady asked the commodore where the sextant was, and he said, "Oh, we don't use that anymore. See that little black box over there? That tells me where I am every hour on the hour within half a degree." So, everybody left the bridge thinking that that little black box was so remarkable, but not relating it to the satellite which was twenty thousand miles up. That was a direct result of the space program which these people did not realize.

Clarke: Well, the oil rigs now in the ocean . . . they rely on satellites to position themselves. This accuracy is now down to a few meters. Much oceanography and oil prospecting couldn't be done without satellite technology.

Downing: It sounds like it might be a public relations problem.

Clarke: It is entirely a public relations problem. We don't seem to know how to handle this. I mentioned my own attempt a few years ago to put all these ideas across to the public and how I spent a lot of money and time on it, but they just didn't want to listen at that time.

Fuqua: I think, Mr. Clarke, you hit the nail on the head, that sometimes the public does not want to listen. I think the Congress has been responsive to scientific development, as in the Manhattan project, and the space program, and the establishment of the National Science Foundation and other science-related work. The problem I think we're concerned about is getting the people to support us so that we can support these programs.

One final question: you mentioned briefly the large space telescope in your testimony. We may have that under consideration before this committee later this year, and maybe next year. Could you just elaborate very briefly on how that would help contribute to the solution of human needs?

Clarke: Astronomy is a science which is in many ways a sort of cutting edge opening up all sorts of basic scientific discoveries. Many of the discoveries of astronomy are very esoteric, but even the esoteric ones do have an extraordinary philosophical appeal, because people are interested about where we're going, where we come from, religious problems. So, there are two ways you can deal with astronomy. One is from the religious-philosophical angle. I mentioned these nonsensical books about visitors from space, and even though they're rubbish, they do show that there is a tremendous public interest in this subject. The other thing is that astronomy has discovered so many things which maybe a generation later, maybe fifty years later, maybe one hundred years later, have been of enormous practical use. The idea of atomic energy came originally from astronomical discoveries; for example, where does the sun get all its energy? This was one of the first clues that thermonuclear reactions could take place.

Unfortunately, many of these points are very hard to put across because they do involve very esoteric scientific ideas. Again, it's an educational problem.

But astronomy does have a beauty about it, and one thing the telescopes do give is beautiful pictures. Incidentally, Skylab produced some of the best examples of modern art in its computer-enhanced displays of the sun. Maybe you would like to get some of the big art museums to have displays of purely scientific photographs —beautiful stuff, which appeals to everybody emotionally. That's why it's such a pity that those big, glossy magazines are no longer in business, because that sort of stuff just doesn't get across to the public.

Downing: How much do you think the cartoon strip Buck Rogers influenced the space program? When I was a child I read in the cartoon strip everything which is happening now.

Clarke: It did have a great influence, I think on the whole a positive influence. It prepared a whole generation for what is happening; although, of course, there is a negative reaction also, because I know at one point people said, "Oh, all that Buck Rogers stuff." You can't have it both ways.

But I think cartoon strips and science fiction generally were helpful—I'm sure we would not have had men on the moon if it had not been for Wells and Verne and the people who write about this and made people think about it. I'm rather proud of the fact that I know several astronauts who became astronauts through reading my books. I feel a considerable responsibility for this myself.

Fuqua: Mr. Clarke, we want to thank you very much for taking your time and sharing both your time and thoughts with us today. You have contributed greatly and we thank you very much for joining us this morning.

Clarke: Thank you very much, Mr. Chairman. I certainly enjoyed speaking to you all.

24

The Second Century of the Telephone

Alexander Graham Bell invented the telephone in Boston on 10 March 1876, and exactly one hundred years later the world's largest corporation, American Telegraph and Telephone, arranged two days of appropriate celebrations at the Massachusetts Institute of Technology. Dr. Jerome Wiesner, provost of M.I.T., invited me to give the concluding address, which was certainly, to coin a phrase, an offer I couldn't refuse.

The visit was memorable in many ways; for the hospitality of Gloria and Marvin Minsky, and the wonders of Dr. Minsky's Artificial Intelligence Laboratory (unfortunately, the famous "Space War" program was on the blink when I was there); for reunions with old friends like John Pierce and Claude Shannon; but especially for the rare privilege of sharing breakfast in his private apartment with one of the greatest living Americans, Dr. Edwin Land—a David just about to sally forth into battle against that corporate Goliath, Eastman Kodak, in defence of his SX-70 camera. I will also long remember Dr. Wiesner's brilliant one-up-manship at the opening session. He mounted the podium of the Kresge Auditorium wearing an elaborate brace around his neck, and spoke with obvious discomfort until he had gained his audience's complete sympathy. Then he casually removed the entire brace and finished his introduction with no signs of distress whatsoever. I consider this almost as good as my own ploy when losing at table tennis; then I am liable to fumble inside my shirt and say apologetically: "Excuse me for a moment while I adjust my pacemaker."

The following week marked another anniversary, for on 16 March 1926, Dr. Robert Goddard made the first flight of a liquid-propelled rocket, near Worcester, Massachusetts. My good friend and long-time Washington host, Fred Durant (Assistant Director, Astronautics, of the new National Air and Space Museum), had arranged appropriate ceremonies and speeches at the Smithsonian, culminating in a dinner in unique surroundings. Even the best of food is liable to take second place, when out of the corner of your eye, you can see the Hope Diamond glittering behind its protective glass.

Yet, in my biassed opinion, that famous jewel is far surpassed in beauty by the other residents of the Museum of Natural History's Gem Room—the much larger and more colorful sapphires and rubies from Sri Lanka. I'd gladly swap the Hope for any of them.

Nevertheless, as the room was full of astronauts and NASA top brass, I did go round looking for volunteers to help me steal it, to get a little money for the beleaguered space program. Whereupon Fred Durant made a memorable comment: "Do you realise," he said loftily, waving his arm around the entire glittering circumference of the Gem Room, "that *all* this collection couldn't pay for my latest exhibit?"

Incredibly, he was right; his remark was not (entirely) a case of that well-known phenomenon, museum director's sour grapes. For Fred had just installed in the great hall of the National Air and Space Museum what must surely be the most valuable single exhibit anywhere on earth: the complete Skylab *backup* (not mockup), just as it would have flown, with everything inside it except the food supplies and other expendables. At a guess, between a quarter and a half *billion* dollars worth of hardware. Anyway, enough to buy at least a hundred Hope diamonds—even assuming that there was no discount for quantity.

The opportunity you have so kindly given me to appear at this convocation allows me to repay a debt of more than thirty years' standing. Back in 1943, as an extremely callow officer in the Royal Air Force, I was given a mysterious assignment to a fog-shrouded airfield at the southwestern tip of England. It turned out that I was to work with an eccentric group of Americans from something called the Radiation Laboratory of the Massachusetts Institute of Technology. They were led by a bright young physicist named Luis Alvarez, who had invented a radar device that, for a change, did something useful. It could bring down an aircraft in *one* piece, instead of several.

The pioneering struggles of the GCA "talk-down" system, and our attempts to convince skeptical pilots that we knew exactly where they were (even when we didn't) you'll find in my only *non*-science-fiction novel[1] ... though *Glide Path* certainly would have been science fiction had it appeared at any time before 1940.

I mention this incident from the palaeoelectronic era for two reasons. Luis's brainchild provided me with the peaceful environment, totally insulated from all the nasty bombings and invasions happening elsewhere, which allowed me to work out the principles of communications

1. *Glide Path* (New York: Harcourt Brace Jovanovich, 1963). See also "You're on the Glide Path—I Think," *The Aeroplane,* September 23, 1949. Reprinted in *IEEE Transactions on Aerospace and Navigational Electronics,* Vol. ANE-10, No. 2 (June 1963).
There is an echo of GCA in my 1945 communications satellite paper. The S-band search radar had a cosecant-squared polar diagram to increase field strength at slant range. I realised that comsats would face a similar problem and therefore suggested the use of a "nonuniform radiator." As the war was still on and my manuscript had to be vetted by RAF security, this was sailing as close to the wind as I dared.

satellites in the spring of 1945. So comsats lie on one of the many roads that lead back to M.I.T.

My second reason for mentioning the primitive technology of a third of a century ago is that it gives a baseline for extrapolation into the future. I can still recall my amazement at the number of vacuum tubes (remember them?) in the GCA Mark 1. It came to the unbelievable total of almost one thousand.

I would have laughed scornfully if some crazy science-fiction writer had predicted that one day every engineer would carry at his hip not a Colt .45 but an H/P 45, a mere handful a *dozen times as complex* as our Mark 1. The explosion in complexity, and the implosion in size, are two of the main parameters determining future communications technology.

Now, in discussing that technology, I'm acutely aware of the fact that I'm competing with people—many of them right here—who have made it their major concern for decades. It's inevitable, therefore, that in the course of this address I'll invent the wheel not once, but several times. My main hope is that *my* wheels will be of novel and interesting shapes, unlike the boringly circular ones produced by all the professionals.

But first, as the commercials say, a few general principles. It's probably true that in communications technology anything that can be conceived, and that does not violate natural laws, can be realised in practice. We may not be able to do it right now, owing to ignorance or economics, but those barriers are liable to be breached with remarkable speed.

For man is the communicating animal; he demands news, information, entertainment, almost as much as food. In fact, as a functioning human being, he can survive much longer without food—even without water!—than without information, as experiments in sensory deprivation have shown. This is a truly astonishing fact; one could construct a whole philosophy around it. (Don't worry—I won't try.)

So any major advance in communications capability comes into widespread use just as soon as it is practicable. Often sooner; the public can't wait for "state of the art" to settle down. Remember the first clumsy phonographs, radios, tape recorders. And would you believe the date of the first music broadcast? It was barely a year after the invention of the telephone! On 2 April 1877 a "telegraphic harmony" apparatus in Philadelphia sent "Yankee Doodle" to sixteen loudspeakers—well, softspeakers—in New York's Steinway Hall. Alexander Graham Bell was in the audience, and one would like to know if he complimented the promoter, his now forgotten rival, Elisha Gray, who got to the Patent Office just those fatal few hours too late.[2]

Now, the telephone is a very simple device, which even the nineteenth century could readily mass produce. In fact, one derivative of the carbon microphone must be near the absolute zero of technological complexity. You can make a working—though hardly hi-fi—microphone out of three carpenter's nails, one laid across the other two, to form a letter H.

The extraordinary—nay, magical—simplicity of the telephone allowed it to spread over the world with astonishing speed. When we consider the very much more complex devices of the future, is it reasonable to suppose that they too will eventually become features of every home, every office?

Since no holds are barred, what about telepathy? Well, I don't believe in telepathy, but I don't *dis*believe in it either. Certainly some form of electronically assisted mental linkage seems plausible; in fact, this has already been achieved in a very crude form, between men and computers, through monitoring of brain waves. However, I find that *my* mental processes are so incoherent, even when I try to focus and organise them, that I should be

---

2. *The New York Times*, May 11, 1975.

very sorry for anyone at the receiving end. Our super-human successors, if any, may be able to cope; indeed, the development of the right technology might force such an evolutionary advance. Perhaps the best that *we* could manage would be the sharing of emotional states, and not the higher intellectual processes. So radio-assisted telepathy might merely lead to some interesting new vices—admittedly, a long-felt want.

Let's stick, therefore, to the recognised sense channels, of which sound and sight are by far the most important. Although one day we will presumably develop transducers for all the senses, just because they are there, I suspect that the law of diminishing returns will set in rather rapidly after the "feelies" and "smellies." These may have some limited applications for entertainment pur-poses, as anyone who was pulverized by the movie *Earthquake* may agree. (Personally, I'm looking forward to the epic *Nova,* in which the theatre's heating system is turned on full blast in the final reel.)

The basic ingredients of the ideal communications device are, therefore, already in common use even today. The standard computer console, with keyboard and visual display, plus hi-fi sound and TV camera, will do very nicely. Through such an instrument (for which I've coined the ugly but perhaps unavoidable name "comsole"—communications console)[3] one could have face-to-face interaction with anyone, anywhere on earth, and send or receive any type of information. I think most of us would settle for this, but there are some other possibilities to consider.

For example; what about *verbal* inputs? Do we really need a keyboard?

I'm sure the answer is yes. We want to be able to type out messages, look at them, and edit them before trans-mission. We need keyboard inputs for privacy, and

---

3. *Imperial Earth* (New York: Harcourt Brace Jovanovich, 1975).

quietness. A *reliable* voice-recognition system, capable of coping with accents, hangovers, ill-fitting dentures,[4] and the "human error" that my late friend HAL complained about, represents something many orders of magnitude more complex than a simple alpha-numeric keyboard. It would be a device with capabilities, in a limited area, at least as good as those of a human brain.

Yet assuming that the curve of the last few decades can be extrapolated, this will certainly be available sometime in the next century. Though most of us will still be tapping out numbers in 2001, I've little real doubt that well before 2076 you will simply say to your comsole, "Get me Bill Smith." Or if you *do* say, "Get me 212–555–5512," it will answer, "Surely you mean 212–555–5521." And it will be quite right.

Now, a machine with this sort of capability—a robot secretary, in effect—could be quite expensive. *It doesn't matter.* We who are living in an economic singularity—if not a fiscal black hole—have forgotten what most of history must be like.

Contrary to the edicts of Madison Avenue, the time will come when it won't be necessary to trade in last year's model. Eventually, everything reaches its technological plateau, and thereafter the only changes are in matters of style. This is obvious when you look at such familiar domestic objects as chairs, beds, tables, knives, forks. Oh, you can make them of plastic or fiberglass or whatever, but the basic design rarely alters.

It took a few thousand years to reach these particular plateaus; things happen more quickly nowadays even for much more complex devices. The bicycle took about a century; radio receivers half that time. This is not to deny that marginal improvements will go on indefinitely,

---

4. No joke, this. My friends at Time-Life once told me that when they edited Churchill's war memoirs they found countless statistical errors. The trouble was traced to the great man's habit of dentureless dictation.

but after a while all further changes are icing on a perfectly palatable cake. You may be surprised to learn that there are electrical devices that have been giving satisfactory service for half a century or more. The other day someone found an Edison carbon filament lamp that has apparently never been switched off since it was installed. And until recently there were sections of Atlantic telegraph cable that had been in service for a full century!

Now, it's hard to see how a properly designed and constructed solid-state device can ever wear out. It should have something like the working life of a diamond, which is adequate for most practical purposes. So when we reach this state of affairs, it would be worth investing more in a multipurpose home communications device than an automobile. It would be handed on from one generation to the next, as was once the case with a good watch.

It has been obvious for a very long time that such audio-visual devices could complete the revolution started by the telephone. As I have written, we are already approaching the point when it will be feasible—not necessarily desirable—for those engaged in what are quaintly called "white-collar" jobs to do perhaps 95 percent of their work without leaving home. Many years ago I coined the slogan: "Don't commute—communicate!" Apart from the savings in travel time (the *real* reason I became a writer is that I refuse to spend more than thirty seconds moving from home to office) there would be astronomical economies in power and raw materials. Compare the amount of hardware in communications systems, as opposed to railroads, highways, and airlines. And the number of kilowatt hours you expend on the shortest journey would power several lifetimes of chatter between the remotest ends of the earth.

Obviously, the home comsole would handle most of today's first-class mail; messages would be stored in its memory waiting for you to press the playback key whenever you felt like it. Then you would type out the answer or, alternatively, call up the other party for a face-to-face chat.

Fine, but at once we have a serious problem—the already annoying matter of time zones. They are going to become quite intolerable in the electronic global village, where we are all neighbours, but a third of us are asleep at any given moment. The other day I was awakened at 4 A.M. by the London *Daily Express,* which had subtracted five and one-half hours instead of adding them. I don't know what I said, but I doubt if my views on the Loch Ness monster were printable.

The railroads and the telegraph made time zones inevitable in the nineteenth century; the global telecommunications network of the twenty-first may abolish them. It's been suggested, at least half seriously,[5] that we'll have to establish a common time over the whole planet—whatever inconvenience this may cause to those old-fashioned enough to gear themselves to the day-night cycle.

During the course of the day—whatever *that* may be— you will use the home comsole to call your friends and deal with business, exactly as you use the telephone now, with *this* difference. You'll be able to exchange any amount of tabular, visual, and graphical information. Thus, if you're an author, you'll be able to wave that horrid page-one typo in front of your delinquent editor on Easter Island or wherever he lives. Instead of spending hours hunting for nonexistent parts numbers, engineers will be able to *show* their supplier the broken doohickey from the rotary discombobulator. And we'll all be able to see those old friends of a lifetime, whom we'll never meet again in the flesh.

Which raises an interesting problem. One of the great advantages of Mr. Bell's invention is that you can converse with people *without* their seeing you, or knowing where

5. By Dr. Solomon Golumb.

you are, or who is with you. A great many business deals would never be consummated, or even attempted, over a video circuit; but perhaps they are deals that shouldn't be, anyway.

I am aware that previous attempts to supply vision, such as the Bell picturephone, have hardly been a roaring success. But I feel sure that this is due to cost, the small size of the picture, and the limited service available. No one would have predicted much of a future for the very first "televisors," with their flickering, postage-stamp-sized images. Such technical limitations have a habit of being rather rapidly overcome, and the *large screen, high definition* picturephone-plus is inevitable.

I could certainly do with such a device. For several years Stanley Kubrick has been talking wistfully to me about another space project. But there's an insoluble problem: I won't leave Sri Lanka for more than a couple of weeks a year, and Stanley refuses to get into an airplane. We may both be too old, or too lazy, before the arrival of home comsoles makes another collaboration possible. So the present backwardness of electronics has spared the world another masterpiece.

Clearly, when we do have two-way vision, there will have to be some changes in protocol. You can't *always* pretend to your wife that the camera has broken down again. Incidentally, some of the changes that would be produced in a society totally oriented to telecommunications have been well discussed by a promising writer, in a novel called *The Naked Sun*. The author's full name escapes me at the moment, but I believe the first name is Isaac.

The possibilities of the comsole as an entertainment and information device are virtually unlimited; some of them, of course, are just becoming available, as an adjunct to the various TV subscription services. At any moment one should be able to call up all the news headlines on the screen, and expand any of particular interest into a complete story at several levels of thoroughness, all the way, let us say, from the *Daily News* to *The New York Times*. I hate to think of the hours I have wasted listening to *radio* news bulletins for some item that never turned up. Nothing is more frustrating—as will be confirmed by any Englishman touring the United States during a Test Match, or any American in England during the World Series (how did it get that ridiculous name?). For the first time, it will be possible to have a news service with immediacy, selectivity, *and* thoroughness.

The electronic newspaper, apart from all its other merits, will also have two gigantic ecological plusses. It will save whole forests for posterity; and it will halve the cost of garbage collection. This alone might be enough to justify it, and to pay for it.

Like many of my generation, I became a news addict during the Second World War. Even now, it takes a definite effort of will for me *not* to switch on the hourly news summaries, and with a truly global service one could spend every waking minute monitoring the amusing, crazy, interesting, and tragic things that go on around this planet. I can foresee the rise of even more virulent forms of news addiction, resulting in the evolution of a class of people who can't bear to miss anything that's happening, anywhere, and spend their waking hours glued to the comsole. I've even coined a name for them— infomaniacs.

Continuing in this vein, I used to think how nice it would be to have access, in one's own home, to all the books and printed matter, all the recordings and movies, all the visual arts of mankind. But would not many of us be completely overwhelmed by such an embarrassment of riches, and solve the impossible problem of selection by selecting nothing? Every day I sneak guiltily past my set of the Great Books of the Western World, most of which I've never even opened. . . . What would it *really* be like to have the Library of Congress—*all* the world's great libraries—at your fingertips? Assuming, of course, that your fingertips were sufficiently educated to handle the problem of indexing and retrieval. I speak with some feeling on this subject, because for a couple of years I had the job of classifying and indexing everything published in the physical sciences, in all languages. If you can't find what you're looking for in *Physics Abstracts* for 1949–1951, you'll know whom to blame.

With the latest techniques, it would be possible to put the whole of human knowledge into a shoe box. The problem, of course, is to get it out again; anything misfiled would be irretrievably lost. Another problem is to decide whether we mass produce the shoe boxes, so that every family has one, or whether we have a central shoe box linked to the home with wide-band communications.

Probably we'll have both, and there are also some interesting compromises. Years ago I invented something that I christened, believe it or not, the "Micropaedia Britannica" (recently I was able to tell Mortimer Adler that I'd thought of the name first). But my "Micropaedia" would be a box about the size of an ordinary hard-cover book, with a display screen and alpha-numeric keyboard. It would contain, in text and pictures, *at least* as much material as a large encyclopaedia plus dictionary. However, the main point of the electronic "Britannica" would not be its compactness, but the fact that, every few months, you could plug it in, dial a number, and have it updated overnight. Think of the saving in wood pulp and transportation that *this* implies!

It is usually assumed that the comsole would have a flat TV-type screen, which would appear to be all that is necessary for most communications purposes. But the ultimate in face-to-face electronic confrontation would be when you could not tell, without touching, whether or not the other person was physically present; he or she would appear as a perfect 3-D projection. This no longer appears fantastic, now that we have seen holographic displays that are quite indistinguishable from reality. So I am sure that this will be achieved someday; I am not sure how badly we need it.

What *could* be done, even with current techniques, is to provide 3-D, or at least wide-screen Cinerama-type, pictures for a single person at a time. This would need merely a small viewing booth and some clever optics, and it could provide the basis for a valuable educational-entertainment tool, as Dennis Gabor has suggested.[6] But it could also give rise to a new industry: personalized television safaris. When you can have a high-quality cinema display in your own home, there will certainly be global audiences for specialized programs with instant feedback from viewer to cameraman. How nice to be able to make a trip up the Amazon, with a few dozen unknown friends scattered over the world, with perfect sound and vision, being able to ask your guide questions, suggest detours, request closeups of interesting plants or animals—in fact, sharing everything except the mosquitoes and the heat!

It has been suggested that this sort of technology might ultimately lead to a world in which no one ever bothered to leave home. The classic treatment of this theme is, of course, E. M. Forster's *The Machine Stops*, written seventy years ago as a counterblast to H. G. Wells.

Yet I don't regard this sort of pathological, sedentary society as very likely. "Telesafaris" might have just the opposite effect. The customers would, sooner or later, be inspired to visit the places that really appealed to them, mosquitoes notwithstanding. Improved communications will promote travel for *pleasure;* and the sooner we get rid of the other kind, the better.

So far, I have been talking about the communications devices in the home and the office. But in the last few decades we have seen the telephone begin to lose its metal umbilical cord, and this process will accelerate. The rise of walkie-talkies and citizen's band radios is a portent of the future.

The individual wristwatch telephone through which you can contact anyone, anywhere, will be a mixed

---

6. *The Mature Society,* 1972.

blessing which, nevertheless, very few will be able to reject. In fact, we may not have a choice; it is all too easy to imagine a society in which it is illegal to switch off your receiver, in case the Chairman of the Peoples' Cooperative wants to summon you in a hurry. But let's not ally ourselves with those reactionaries who look only on the *bad* side of every new development. Alexander Graham Bell cannot be blamed for Stalin, once aptly described as "Genghis Khan with a telephone."

It would be an *underestimate* to say that the wristwatch telephone would save tens of thousands of lives a year. Everyone knows of tragedies—car accidents on lonely highways, lost campers, overturned boats, even old people at home—where some means of communication would have made all the difference between life and death. Even a simple emergency SOS system, whereby one pressed a button and sent out a Help! signal, would be enough. This is a possibility of the immediate future; the only real problem and, alas, a serious one, is that of false alarms.

At this point, before I lose all credibility with the hairy-knuckled engineers who have to produce the hardware, I'd better do a once-over-lightly of the electromagnetic spectrum. This is, I think, unique among our natural resources. We've been exploiting it for less than one lifetime, and are now polluting much of it to the very maximum of our ability. But if we stopped using it tomorrow, it would be just as good as new, because the garbage is heading outwards at the speed of light. Too bad this isn't true of the rest of the environment.

Do we have enough available band-width for a billion personal transceivers, even assuming that they aren't all working at once? As far as the home equipment is concerned, there is no problem, at least in communities of any size. The only uncertainty, and a pretty harrowing one to the people who have to make the decisions, is how quickly coaxial cables are going to be replaced by glass fibres, with their millionfold greater communications capability. Incidentally, one of the less glamorous occupations of the future will be mining houses for the rare metal, copper, buried inside them by our rich ancestors. Fortunately, there is no danger that we shall ever run out of silica.

But I would also suggest that optical systems, in the infrared and ultraviolet, have a great future not only for fixed, but even for *mobile,* personal communications. They may take over some of the functions of present-day transistor radios and walkie-talkies—leaving the radio bands free for services which can be provided in no other way. The fact that opticals have only very limited range, owing to atmospheric absorption, can be turned to major advantage. You can use the same frequencies—and *what* a band of frequencies!—millions of times over, as long as you keep your service areas ten or twenty kilometres apart.

It may be objected that light waves won't go round corners, or through walls. Elementary, my dear Watson. We simply have lots of dirt-cheap (because they are made from dirt!) optical-waveguides and light pipes deliberately leaking radiation all over the place. Some would be passive, some active. Some would have very low-powered optical-to-radio transducers in both directions, to save knocking holes in walls, and to get to awkward places. In densely populated communities one would always be in direct or reflected sight of some optical transmitter or repeater. But we must be careful how we use the ultraviolet. People who talked too much might get sunburned.

When you are cycling across Africa, or drifting on a balsa-wood raft across the Pacific, you will of course still have to use the radio frequencies, say, the one to ten thousand megahertz bands, which can accommodate at least ten million voice circuits. This number can be multiplied many times by skillful use of satellite technology. I can envisage an earth-embracing halo of low-altitude, low-powered radio satellites, switching frequencies continually so that they provide the desired coverage in given geographical regions. And NASA has recently published a most exciting report on the use of the very large (kilometre-square!) antennas we will soon be able to construct in space.[7] These would permit the simultaneous use of myriads of very narrow beams, which could be focussed on individual subscribers, carrying receivers which could be mass produced for about ten dollars. I rather suspect that our long-awaited personal transceiver will be an adaptive, radio-optical hybrid, actively hunting the electromagnetic spectrum in search of incoming signals addressed to it.

Now, the invariably forgotten accessory of the wristwatch telephone is the wristwatch telephone *directory*. Considering the bulk of that volume for even a modest-sized city, this means that our personal transceivers will require some sophisticated information-retrieval circuits, and a memory to hold the few hundred most-used numbers. So we may be forced, rather quickly, to go the whole way, and combine in a single highly portable unit not only communications equipment but something like today's pocket calculators, plus data banks, plus information-processing circuits. It would be a constant companion, serving much the same purpose as a human secretary. In a recent novel I called it a "Minisec."[8] In fact, as electronic intelligence develops, it would provide more and more services, finally developing a personality of its own, to a degree which may be unimaginable today.

Except, of course, by science-fiction writers. In his brilliant novel, *The Futurological Congress*, Stanislaw Lem gives a nightmare cameo which I can't get out of my mind. He describes a group of women sitting in complete silence, while their handbag computers gossip happily to one another.

---

7. Aerospace Corporation Report, *Potential Space System Contributions in the Next Twenty-Five Years*, 1975. For the Summary, see volume 2 of the House of Representatives Subcommittee on Space Science and Applications, *Future Space Programs, 1975*.

8. In *Imperial Earth*.

The marvellous toys that we have been discussing will simply remain toys, unless we use them constructively and creatively. Now, toys are all right in the proper place; in fact, they are an essential part of any childhood. But they should not become mere distractions, or ways of drugging the mind to avoid reality.

We have all seen unbuttoned beer-bellies slumped in front of the TV set, and transistorised morons twitching down the street, puppets controlled by invisible disk jockeys. These are not the highest representatives of our culture; but, tragically, they may be typical of the near future. As we evolve a society oriented towards information, and move away from one based primarily on manufacture and transportation, there will be millions who cannot adapt to the change. We may have no alternative but to use the lower electronic arts to keep them in a state of drugged placidity.

For in the world of the future, the sort of mindless labor that has occupied 99 percent of mankind for much more than 99 percent of its existence will of course be largely taken over by machines. Yet most people are bored to death without work, even work they don't like. In a workless world, therefore, only the highly educated will be able to flourish, or perhaps even to survive. The rest are likely to destroy themselves and their environment out of sheer frustration. This is no vision of the distant future; it is already happening, most of all in the decaying cities.

So perhaps we should not despise TV soap operas if, during the turbulent transition period between our culture and real civilisation, they serve as yet another opiate for the masses. *This* drug, at any rate, is cheap and harmless, serving to kill time—for those many people who like it better dead.

When we look at the manifold problems of our age, it is clear that the most fundamental one—from which almost all others stem—is that of ignorance. And ignorance can be banished only by communication, in the widest meaning of the word. The best educational arrangement, someone once remarked, consists of a log with a teacher at one end and a pupil at the other. Unfortunately, there are no longer enough good teachers, and probably not enough logs, to go around.

Now, one thing that electronics can do rather well is to multiply teachers, as NASA's ATS-6 satellite broadcasting educational programs to several thousand villages in India is demonstrating. Thanks to the extraordinary generosity of the Indian Space Research Organisation, which flew in six engineers and half a ton of equipment, I have a five-metre parabolic antenna on the roof of my Colombo house, now renamed Jodrell Bank East. Since the experiment started on 1 August 1975, I have thus been in the curious position of having the only TV set in the country. It's been fascinating to watch the programs. Even though I don't understand the language, the messages of family planning, hygiene, agricultural techniques, and national unity come across loud and clear.

Though it is impossible to put a value on such things, I believe that the cost of this experiment will be trivial compared with the benefits. And the ground segment is remarkably cheap, in terms of its coverage. Would you believe four thousand people round one TV set? Or a three-metre diameter village antenna—made of *dried mud?*

Of course, there are some critics, as reported recently by Dr. Yash Pal, the able and energetic director of the Indian Space Application Centre.

In the drawing rooms of large cities you meet many people who are concerned about the damage one is going to cause to the integrity of rural India by exposing her to the world outside. After they have lectured you about the dangers of corrupting this innocent, beautiful mass of humanity, they usually turn around and ask: "Well, now that we have a satellite, when are we going to see some American programs?" Of course, they themselves are immune to cultural domination or foreign influences.[9]

I'm afraid that cocktail-party intellectuals are the same everywhere. Because *we* frequently suffer from the modern scourge of information pollution, we find it hard to imagine its even deadlier opposite—information starvation.

Unfortunately, on 31 July 1976 the one-year experiment will end; ATS-6 will crawl back along the equator and return to the United States. Originally, it was hoped that two satellites would be launched; last summer I saw the three-quarters completed ATS-7 sitting mothballed at the Fairchild plant. No one could raise the ten million necessary to finish it, or hijack one of the Air Force's numerous Titan 3-C's to get it into orbit.

And so in a few months' time, millions of people who have had a window opened on marvellous new worlds of culture and education will have it slammed shut in their faces again. There will be some heartrending scenes in the villages, when the cry goes up, however unfairly, "The Americans have stolen our satellite!" Useless to explain, as the frustrated viewers start to refill their six to nine time slot with baby-making, that it was only through the initiative and generosity of the United States that the satellite was loaned in the first place. The Ugly American will have struck again.

Yet I hope that this noble experiment is just the curtain raiser to a truly global educational satellite system. Its cost would be one or two dollars per student per *year*. There could be few better investments in the future health, happiness, and peace of mankind.

I don't wish to get too much involved in the potential— still less the politics—of communications satellites, be-

---

9. "Some Lessons During the Setting Up of SITE." (Talk at UN/UNESCO Regional Seminar on Satellite Broadcasting Systems for Education and Development, Mexico City, September 2–11, 1975.)

cause they can take care of themselves. The world investment in satellites and ground stations now exceeds a billion dollars, and is increasing almost explosively. After years of delay and dithering, the United States is at last establishing *domestic* satellite systems; the U.S.S.R. has had one for almost a decade. At first, the Soviet network employed *non*synchronous satellites, moving in an elongated orbit that took them high over Russia for a few hours of every day. However, they have now seen the overwhelming advantages of stationary orbits, and several of their comsats are currently fixed above the Indian Ocean. Some are designed for TV relaying to remote parts of the Soviet Union, and I've gently hinted to my friends in Moscow that perhaps *they* could fill the breach when ATS-6 goes home.

We are now in the early stages of a battle for the mind—or at least the eyes and ears—of the human race, a battle which will be fought thirty-six thousand kilometres above the equator. The preliminary skirmishes have already taken place at the United Nations, where there have been determined attempts by some countries to limit the use of satellites which can beam programs from space directly into the home, thus bypassing the national networks. Guess who is scared?

As a matter of fact, I tried to frighten the United States with satellites myself, back in 1960, when I published a story in *Playboy*[10] about a Chinese plot to brainwash innocent Americans with pornographic TV programs. Perhaps "frightened" is not the correct verb, and in these permissive days such an idea sounds positively old-fashioned. But in 1960 the first regular comsat service was still five years in the future, and this seemed a good gambit for attracting attention to its possibilities.

Fortunately, in this area there is an excellent record of international cooperation. Even countries who hate each

---

10. "I Remember Babylon," *Playboy,* May 1960. Reprinted in *Tales of Ten Worlds.*

other's guts work together through the International Tele-communications Union, which sets limits to powers and assigns frequencies. Eventually, some kind of consensus will emerge which will avoid the worst abuses.

I would like to end with some thoughts on the wider future of communications—communications beyond the earth. And here we face an extraordinary paradox, which in the centuries to come may have profound political and cultural implications. For the whole of human history, up to that moment one hundred years ago, it was impossible for two persons more than a few metres apart to interact in real time. The abolition of that apparently fundamental barrier was one of technology's supreme triumphs; today we take it for granted that men can converse with each other, and even see each other, wherever they may be. Generations will live and die, always with this godlike power at their fingertips. Yet this superb achievement will be ephemeral; before the next hundred years have passed our hard-won victory over space will have been lost, never to be regained.

On the Apollo voyages, for the first time, men travelled more than a light-second away from the earth. The result-ing two-and-a-half-second roundtrip delay was surpris-ingly unobtrusive, but only because of the dramatic nature of the messages—and the discipline of the speakers. I doubt if the average person will have the self-control to talk comfortably with anyone on the moon.

And beyond the moon, of course, it will be impossible. We will never be able to converse with friends on Mars, even though we can easily exchange any amount of infor-mation with them. It will take at least three minutes to get there, and another three minutes to receive a reply.[11]

11. See J. J. Coupling, "Don't Write; Telegraph," *Astound-ing Science Fiction,* March 1952. Mr. Coupling was a promising science-fiction writer whose output was sadly limited by the activities of his alter ego, Dr. John Pierce, director of Com-munications Research at Bell Labs. See also the chapter "The Politics of Time and Space," in *Imperial Earth.*

Anyone who considers that this is never likely to be of much practical importance is taking a very short-sighted view. It has now been demonstrated, beyond reasonable doubt, that in the course of the next century we could occupy the entire solar system. The resources in energy and material are there; the unknowns are the motivation and our probability of survival, which may indeed depend upon the rate with which we get our eggs out of this one fragile planetary basket.

We would not be here, talking about the future, unless we were optimists. And in that case we must *assume* that eventually very large populations will be living far from Earth—light-minutes and light-hours away, even if we only colonise the inner solar system. However, Freeman Dyson has argued with great eloquence[12] that planets aren't important, and the real action will be in the cloud of comets out beyond Pluto, a light-*day* or more from Earth. And looking further afield, it is now widely realised that there are no *fundamental* scientific obstacles even to interstellar travel.[13] Though Dr. Purcell once rashly remarked that starships should stay on the cereal boxes, where they belong, that's exactly where moonships were, only thirty years ago.

So the finite velocity of light will, inevitably, divide the human race once more into scattered communities, sundered by barriers of space and time. We will be as one with our remote ancestors, who lived in a world of immense and often insuperable distances, for we are moving out into a universe vaster than all their dreams.

---

12. "The World, the Flesh and the Devil," Third J. D. Bernal Lecture, Birkbeck College, London, 1972. Now available as Appendix D to *Communications With Extraterrestrial Intelligence,* edited by Carl Sagan (Boston: MIT Press, 1973).

13. See, for example, the "Interstellar Studies" issues of the *Journal of the British Interplanetary Society.* Just forty years ago, amid general incredulity, *J.B.I.S.* started to publish studies of vehicles which could carry men to the moon. This is where we came in.

But it is, surely, not an empty universe. No discussion of communications and the future would be complete without reference to the most exciting possibility of all: communications with extraterrestrial intelligence. The galaxy must be an absolute Babel of conversation, and it is surely only a matter of time before we can hear the neighbours. They already know about us, for our sphere of detectable radio signals is now scores of light-years across. Perhaps even more to the point—and more likely to bring the precinct cops hurrying here as fast as their paddy wagon can travel—is that fact that several microsecond-thick shells of X-rays are already more than ten light-years out from Earth, announcing to the universe that, somewhere, juvenile delinquents are detonating atom bombs.

Plausible arguments suggest that our best bet for interstellar eavesdropping would be in the thousand-megahertz, or thirty-centimetre, region. The NASA/Stanford/Ames Project Cyclops report, which proposed an array of several hundred large radio telescopes for such a search, recommended a specific band about two hundred megahertz wide—that lying between the hydrogen line (1420 MHz) and the lowest OH line (1662 MHz). Dr. Bernard Oliver, who directed the Cyclops study, has waxed poetic about the appropriateness of *our* type of life seeking its kind in the band lying between the disassociation products of water—the "water hole."[14]

Unfortunately, we may be about to pollute the water hole so badly that it will be useless to radio astronomers. The proposed Marisat and Navstar satellites will be dunked right in the middle of it, radiating so powerfully that they would completely saturate any Cyclops-type array. Barney Oliver tells me: "Since the Cyclops study, additional reasons have become apparent for expecting

---

14. Project Cyclops: A Design Study of a System for Detecting Extraterrestrial Intelligent Life (NASA/Ames CR 114445).

the water hole to be our contact with the mainstream of life in the galaxy. The thought that we, through our ignorance, may blind ourselves to such contact and condemn the human race to isolation appalls us."

I hope that the next World Administrative Radio Conference, when it meets in 1979, will take a stand on this matter. The conflict of interest between the radio astronomers and the communications engineers will get more and more insoluble until, as I suggested many years ago,[15] we move the astronomers to the quietest place in the solar system—the centre of the lunar farside, where they will be shielded from the radio racket of earth by thirty-five hundred kilometres of solid rock. But *that* answer will hardly be available before the next century.

Whatever the difficulties and problems, the search for extraterrestrial signals will continue. Some scientists fear that it will not succeed; others fear that it *will*. It may already have succeeded, but we don't yet know it. Even if the pulsars *are* neutron stars—so what? They may still be artificial beacons, all broadcasting essentially the same message: "Last stop for gas this side of Andromeda."

More seriously, if the decades and the centuries pass with no indication that there is intelligent life elsewhere in the universe, the long-term effects on human philosophy will be profound, and may be disastrous. Better to have neighbours we don't like than to be utterly alone. For that cosmic loneliness could point to a very depressing conclusion: that intelligence marks an evolutionary dead end. When we consider how well—and how *long*—the sharks and the cockroaches have managed without it, and how badly we are managing *with* it, one cannot help wondering if intelligence is an aberration like the armour of the dinosaurs, dooming its possessors to extinction.

---

15. "The Uses of the Moon," *Harper's,* December 1961. Reprinted in *Voices from the Sky* (New York: Harper & Row, 1965).

No, I don't *really* believe this. Even if the computers we carry on our shoulders are evolutionary accidents, they can now generate their own programs—and set their own goals.

For we can now say, in the widest possible meaning of the phrase, that the purpose of human life is information processing. I have already mentioned the strange fact that men can survive longer without water than without information. And therefore the real value of all the devices we have been discussing is that they have the potential for immensely enriching and enlarging life, by giving us more information to process—up to the maximum number of bits per second that the human brain can absorb.

I am happy, therefore, to have solved one of the great problems the philosophers and theologians have been haggling over for several thousand years. You may, perhaps, feel that this is rather a dusty answer, and that not even the most inspired preacher could ever found a religion upon the slogan: "The purpose of life is information processing." Indeed, you may even retort, "Well, what is the purpose of information processing?"

I'm glad you asked me that.

25

Ayu Bowan!

My attempt to survey the telephonic future is still, to borrow a phrase from the British humourist Beachcomber, by far the only writing I have done this year (mid-1976), and I will be very happy if it remains so. Although my next novel, *The Fountains of Paradise,* is supposed to be delivered on my sixtieth birthday (16 December 1977), and I have been thinking about it for ten years, I have put not a single word on paper. I hope it will not take as long to produce as *Imperial Earth,* which had a gestation period of more than two decades.

But I make no promises, for Sri Lanka seems to have as many distractions as London or New York, as well as some unique ones of its own. For example, I am typing this with some difficulty, owing to the presence in my lap of Miss Kong, a very pretty baby monkey (purple-faced langur) who suffers from the unshakable conviction, based on the flimsiest of evidence, that I am her next of kin. This strange delusion causes her to cry continuously except when I take pity on her; then she is perfectly happy to cling to me for hours without making a sound. (Luckily she is lapbroken, if not completely housebroken.) She is liable to chase away, with awesome displays of jealous rage, other members of my establishment, up to one hundred times her weight.

Fortunately for Miss Kong, my two beloved German shepherds, Sputnik and Rex, regard her with tolerant amusement. If they considered her a serious rival for my affections, her emotional problems would soon be over.

The largest single consumer of my time is that welcome

but dreaded figure, the postman, who is likely to dump fifty pieces of mail on me in a single delivery. Now that I am at the receiving end myself, I feel guilty about my own extensive impositions upon such amiable writers as Lord Dunsany and C. S. Lewis, who replied to me at some length (this correspondence is now being edited for publication by Willis Conover). For many years, possibly as a result of my civil-service background, I was a compulsive letter-answerer, giving a personal reply within twenty-four hours whenever this was humanly possible.

No longer. About five years ago I was compelled to design a sort of general-purpose mimeographed reply which covered 90 percent of the queries I received; without this, I should have been doomed, and certainly my last two novels would never have been written. Here is my lifesaver; for the benefit of other harassed authors, I waive all copyright.

As I now receive several thousand items of mail a year, it is not possible for me to answer you personally. Moreover, it is sometimes many months before letters reach me. This reply is designed to deal with about 90 percent of the questions I am asked. I hope you will understand the need for it, and I thank you for your interest in my work.

BIOGRAPHY: Biographical details will be found in the standard references; for example, *Who's Who, Contemporary Authors, Contemporary Novelists, Dictionary of International Biography, Celebrity Register, Britannica 3*. For Ceylon background see the accounts of my underwater activities there, especially "The Treasure of the Great Reef," *New Yorker* "Profile," 9 August 1969.

BIBLIOGRAPHY: Now about fifty books; see above references, or the list in any recent edition.

RIGHTS: All editorial queries should be addressed to:

> David Higham Associates
> 5-8 Lower John Street
> Golden Square
> London WIR 4HA
> (01-437-7888)

> or

> Scott Meredith
> 845 Third Avenue
> New York, N.Y. 10022
> (212/245-5500)

LECTURES: I no longer accept lecture requests outside Sri Lanka.

SPACE: Obviously, I cannot answer queries about space; there are hundreds of books on the subject readily available. For general information in this field, write to the Secretary, The British Interplanetary Society, 12 Bessborough Gardens, London, S.W.1., or to the Department of Public Affairs, NASA, Washington, D.C. 20546.

PHOTOS: I am sorry, but I *cannot* supply photos and autographs. In particular, I cannot autograph and mail back books! Any sent here are donated to local libraries.

UNDER NO CIRCUMSTANCES will I comment on manuscripts. For some of the reasons, see the chapter, "Dear Sir . . ." in *Voices from the Sky.* Nor am I interested in ideas for stories, as I already have far more than I can ever use!

ADVICE TO AUTHORS: The only advice I can give to would-be authors is as follows: Read at least one book a day and write as much as you can. Read the memoirs of authors who interest you. (Somerset Maugham's *A Writer's Notebooks* is a good example.) Correspondence courses, writer's schools, and so on, are probably useful—but all the authors I know were self-taught. There is no substitute for living. As Hemingway remarked: "Writing is not a fulltime occupation."

2001: The answer to all queries on this subject will be found in the novel *2001: A Space Odyssey* (NAL in U.S., Corgi in U.K.); the book *The Lost Worlds of 2001* (NAL in U.S., Sidgwick and Jackson in U.K.); the two concluding essays in *Report on Planet Three* (Harper & Row in U.S., Gollancz in U.K.). Jerome Agel's *The Making of Kubrick's 2001* (NAL) is also a very useful reference, with many photos.

PUBLICITY QUOTES: So many publishers and authors have asked me to comment on books, or to write prefaces, that I am now forced to turn down *all* such requests, no matter how good the cause.

Hopefully, this will head off a few thousand letters at the pass. Yet I fear that very soon it will be impossible to send out even this form; I am seriously thinking of falling

back to the next line of defense—a simple postcard with
the words

MR. CLARKE REGRETS ...

And after *that*—silence.

Personal visitors present even more difficult problems,
because I *like* meeting people, and the occasion arises
about ten times a day. Luckily I am able to insulate myself, in my office on the upper floor, from the flow of customers to my partner Hector Ekanayake's company,
Underwater Safaris, though quite often I get involved in
that activity. Thus when the Apollo 12 team, plus wives
and staff, visited Ceylon after their return from the moon,
I accompanied the party while Hector took them to the
bottom of Trincomalee harbour. However, I sat in the
boat, anxiously counting the number of divers who came
out, and comparing it with the number that went in.
(Later, on a roll of film shot by Pete Conrad, I found a
closeup of Jane Conrad's hand, apparently in the act of
patting a friendly scorpion fish—an encounter which
would certainly be painful, and could be fatal.)

Writers, film and TV personalities, diplomats, photographers, scientists, politicians, journalists, now flow in
a steady stream through from Bandaranaike airport, and
a substantial percentage seems to end up sooner or later
at my house.

The greatest flood of visitors—occasionally rising to
over fifty a day—was provoked by SITE. As mentioned
in Chapter 24, in August 1975, to my delighted astonishment, the Indian Space Research Organisation installed
a satellite ground station on my roof, to receive the beautifully clear programs from ATS-6.

The language problem did not stop viewers sitting for
hours, hypnotised by the miracle of images falling down
from the sky after a journey of more than seventy thou-

sand kilometres. As I was anxious for as many people as possible to see the programs, I (somewhat rashly) issued an open invitation. So at any moment, a bus was liable to arrive from some remote corner of the island and disgorge a load of virginal viewers.

I am able to visit my favourite spot (Chapter 13) for only a few days a year. But now, quite unexpectedly— and literally since I wrote the preceding paragraph!— Serendipity has struck again. While researching a totally different subject, I've discovered a good reason for spending more time on the south coast.

It concerns the great Sanskrit epic, the *Ramayana*. In this 2,200-year-old poem, the demon-king Ravanna kidnaps Sita, wife of Rama, and takes her to his island stronghold of Ceylon. Needless to say, she is ultimately released, after aerial battles involving what look suspiciously like atomic weapons and laser beams.

To heal the wounded, the heroic monkey-general Hanuman is later sent back to India to fetch a medicinal herb found only in the Himalayas. Unfortunately, when he gets to the right mountain he is unable to identify the herb. No problem; he brings the whole mountain back! However, one piece drops off, on the southern tip of Ceylon. The locals believe that this fragment is in fact my favourite bay, for its name in Sinhalese means "there it fell down" (*onna watuna*).

There it fell down. Place names usually have a meaning, though it is often lost in the mists of time. Did something *really* fall down, centuries or millennia ago, at Unawatuna Bay? A meteorite would be the obvious explanation; it must have been a big one for the legend to have lasted down the ages.

And here's another weird coincidence. Little Unawatuna, believe it or not, is the closest point on dry land to the world's greatest gravitational anomaly, a few hundred kilometres out in the Indian Ocean. On the Goddard Space Flight Center's 3-D map of the Earth's Gravimetric Geoid, that strange phenomenon looks like a deep pit[1] into which the whole island of Sri Lanka is about to slide.

Let's put two and two together. A few thousand years ago, a huge object of peculiar density plunged into the Indian Ocean, creating a tradition that is remembered to this day. And it's still there, distorting the earth's gravitational field—Terran Gravitic Anomaly I.

That might make an opening for a pretty good science-fiction movie . . . and an even better ending for this book.

Ayu Bowan.

---

1. One hundred and ten metres below zero reference on the Goddard model (Marsh & Vincent, 1974).

About the Author

ARTHUR C. CLARKE was born at Minehead, Somerset, England, in 1917 and is a graduate of King's College, London, where he obtained First Class Honors in Physics and Mathematics. He is past chairman of the British Interplanetary Society, a member of the Academy of Astronautics, the Royal Astronomical Society, and many other scientific organizations. During the Second World War, as an RAF officer, he was in charge of the first radar talk-down ("GCA") equipment during its experimental trials. His only *non*–science-fiction novel, *Glide Path,* is based on this work.

Author of almost fifty books, some twenty million copies of which have been printed in over thirty languages, he has won numerous awards, including a gold medal of the Franklin Institute for having originated communications satellites in a technical paper published in 1945; the 1961 Kalinga Prize; the Aviation-Space Writers' 1965 prize for the best aerospace reporting of the year for his article on comsats in *Life;* the AAAS-Westinghouse science-writing prize; and the HUGO, NEBULA, and John W. Campbell Awards—all three of which were won by his novel *Rendezvous with Rama.*